TECHNICAL REPORT

General Military Training

Standardization and Reduction Options

Roland J. Yardley, Dulani Woods,
Cesse Cameron Ip, Jerry M. Sollinger

Prepared for the Office of the Secretary of Defense

Approved for public release; distribution unlimited

RAND NATIONAL DEFENSE RESEARCH INSTITUTE

The research described in this report was prepared for the Office of the Secretary of Defense (OSD). The research was conducted within the RAND National Defense Research Institute, a federally funded research and development center sponsored by OSD, the Joint Staff, the Unified Combatant Commands, the Navy, the Marine Corps, the defense agencies, and the defense Intelligence Community under Contract W74V8H-06-C-0002.

Library of Congress Control Number: 2012955339

ISBN: 978-0-8330-6912-2

Published 2012 by the RAND Corporation
1776 Main Street, P.O. Box 2138, Santa Monica, CA 90407-2138
1200 South Hayes Street, Arlington, VA 22202-5050
4570 Fifth Avenue, Suite 600, Pittsburgh, PA 15213-2665
RAND URL: http://www.rand.org/
To order RAND documents or to obtain additional information, contact
Distribution Services: Telephone: (310) 451-7002;
Fax: (310) 451-6915; Email: order@rand.org

Preface

Every uniformed service member—whether Active Component (AC) or Reserve Component (RC)—must complete ancillary or general military training (GMT) requirements prescribed by his or her service. Training requirements vary across the services, both in what is trained and in how long it takes to train, but some requirements tend to be similar. The U.S. Department of Defense has identified a need to reduce cyclic mandatory training (MT) requirements (especially for the RCs), thus reducing the training burden on the services and making the most of available training time.

The RAND National Defense Research Institute (NDRI) was asked to examine the services' mandatory military training requirements, also referred to as *general military* or *common military training* requirements, and examine options to standardize requirements and reduce the training burden. This report responds to that request by providing a common definition of *GMT* and examining both the guidance that directs GMT completion and the services' approaches to conducting GMT. The report presumes some knowledge of the terminology associated with military training. The report should be of interest to the military training community. This report represents an analysis of the service GMT programs through October 2011. It is likely that, in the interim, there have been some changes to the services' GMT programs.

This research was sponsored by the Office of the Assistant Secretary of Defense for Reserve Affairs (OASD RA) and conducted within the Forces and Resources Policy Center of NDRI, a federally funded research and development center sponsored by the Office of the Secretary of Defense, the Joint Staff, the Unified Combatant Commands, the Navy, the Marine Corps, the defense agencies, and the defense Intelligence Community.

For more information on the RAND Forces and Resources Policy Center, see http://www.rand.org/nsrd/ndri/centers/frp.html or contact the director (contact information is provided on the web page).

Contents

Figures

Tables

Summary

Background

All the military services train their members on general military topics, such as the Code of Conduct (CoC) or suicide prevention. Individual services direct some topics, and some are stipulated by the U.S. Department of Defense (DoD). Typically, military members receive instruction on these topics at some designated frequency, e.g., annually. This instruction competes for time and other resources with topics specific to a given service.

General military training (GMT) requirements are the same for the Active Component (AC) and Reserve Component (RC). However, AC personnel have greater availability for training, while RC personnel normally drill one weekend per month and complete a two-week training period during the year. Therefore, although the time it takes to perform mandatory GMT is equal for the AC and RC, the time available for drilling reservists to complete the requirements is less. And, with similar GMT demands, the time required to complete mandatory GMT requirements consumes a larger portion of an RC member's available training time.

DoD wants to reduce cyclic mandatory training (MT) requirements for the RC, thus easing the premobilization training burden on the services and increasing available training time. A goal of DoD's current effort is to optimize available training time and effectiveness and return time to the RC unit commanders.

Purpose

The objective of this research is to analyze current training approaches that the services use in meeting GMT needs, assess the GMT that is performed throughout the services, and identify opportunities to reduce and standardize training to meet GMT needs.

The analytical approach for this study involved six tasks:

- Develop a definition for GMT.
- Identify GMT requirements directed by law and policy (including those directed by DoD, the Army, the Navy, the Marine Corps, and the Air Force).
- Evaluate different methods of delivery of the training.
- Establish a baseline of common requirements.
- Evaluate the requirements.
- Provide recommendations for standardization and reduction.

The services do not have a common definition of *GMT*, nor does a DoD-wide definition exist. We examined the services' definitions of *GMT* and, from those definitions, developed a working definition that the sponsor approved. The approved working definition follows:

> *GMT:* periodic, nonoccupational directed training that provides common knowledge and skills required for all uniformed personnel. Ancillary training or GMT enhances an individual's ability to (be prepared to) perform military duties or activities. An example of GMT is sexual assault awareness and prevention.

RAND researchers identified GMT requirements directed by law and policy, including those directed by DoD, the Army, the Navy, the Marine Corps, and the Air Force. The researchers reviewed laws and DoD and service policies and instructions that relate to GMT. In addition, the research team interviewed service AC and RC subject-matter experts (SMEs), i.e., staff officers responsible for GMT policy, both individually and during an Office of the Secretary of Defense (OSD)–sponsored group session to gain a deeper understanding of the issues and challenges that GMT presents. Service SMEs provided the research team with the current service publications that direct completion of GMT.[1]

We assessed DoD directives and service GMT programs to identify GMT topics that met the working definition. Our assessment of services' training topics and approaches indicate that, although some training topics are common among the services' GMT programs, many topics are unique to each service. Some GMT topics that are mandatory in one service are not mandatory in others. The services also vary with regard to the number of required GMT topics on which uniformed personnel must receive training. Moreover, although DoD guidance directs training requirements that are required by law, many GMT requirements are driven solely by service policy. The services develop and provide GMT independently, and the prescribed delivery methods are different.

The Army, Navy, and Coast Guard GMT program requirements are required training for all uniformed personnel. The Air Force and Marine Corps organize their GMT programs differently:

- The Air Force divides its GMT into four categories (Total Force Awareness Training [TFAT], selected force training, event-driven training, and expeditionary skill training [EST]). TFAT is required for all uniformed personnel; selected force training is targeted to specific groups of personnel (e.g., commanders, supervisors); event-driven training is a requirement triggered by an event, such as in-processing; and EST is associated specifically with contingencies. The Air Force TFAT, selected force, and event-driven training are covered in Appendix E.
- The Marine Corps divides its training into two categories: formal and ancillary. Formal training requirements are those established by Marine Corps orders and directives, and ancillary training requirements are "additional, secondary training requirements that provide information to improve the administrative, social, or cultural aspects of military service" (Marine Corps Reference Publication [MCRP] 3-0A, p. C-2). We cover the Marine Corps formal and ancillary training requirements in Appendix F.

---

1 Many service SMEs with whom we met indicated that the service GMT guidance is in flux and that governing policies and instructions would soon be updated.

The services use various methods to provide GMT. These methods range from traditional, instructor-provided training in a classroom to computer-based training (CBT). The Army, Navy, and Marine Corps prescribe instructor-led training as the primary delivery method. The Air Force's and Coast Guard's primary method is CBT. CBT can be interactive, in which a comprehension test is included in the instruction, or noninteractive (no comprehension test). Some GMT, such as physical fitness or marksmanship, is completed through hands-on training. Some GMT topics are augmented or delivered by video instruction. The preferred method of delivering the training also varies by service and, in some cases, by component within each service.

Findings

We find the following from our research.

Time Devoted to General Military Training

Although GMT in the aggregate requires a substantial number of hours across the services, the amount of time as a percentage of total training time for AC units is not high, constituting less than 1 percent total training time. The burden falls more heavily on some RC units because they must accomplish the same GMT as the AC and have far fewer training hours available. The Army and Marine Corps RC devote from 5 to 8 percent of available training time to GMT.

Standardization

Service GMT programs have been developed independently of one another and are not standardized across services. There appear to be opportunities for standardization, which could yield more-efficient use of time and resources.

Flexibility

All services want GMT that can be tailored to their specific purposes and that can be delivered in different ways to different audiences (such as CBT that could be used for makeup training for individuals who miss regularly scheduled training).

Reducing the General Military Training Burden

Several services have reduced the amount of GMT they require; however, in some cases, they have eliminated training that is still required. Some topics that should be included in service GMT plans are not. Additionally, our review of service GMT programs indicates that some topics that are required for GMT according to the approved working definition are not required training across all services.

Need for Training

Many GMT topics have been long-standing requirements, and these requirements should receive a careful review to determine whether they are still necessary and whether the current frequency requirements remain valid. Any new requirements should be scrutinized carefully.

Single Source of General Military Training Requirements

DoD GMT requirements are spread among many directives, which complicates the services' task of identifying what DoD requires.

Potential Efficiencies

Both the Army and Marine Corps could realize efficiencies, in terms of instructor preparation time and standardized content, by maintaining a site to download instructional materials and instructor guides. These materials could be made available online on their respective learning management systems (LMSs) and would also promote standardization across services and components.

Options

Several ways exist to reduce the GMT burden and to make GMT delivery more efficient. Some services have already implemented options, and these appear to reduce time required for GMT and increase efficiency. However, these options have not been thoroughly analyzed.

Recommendations

In light of these findings, we recommend that DoD or the services take the following actions.

DoD and the services should consider using a DoD-wide accessible site that the services and components could use to download standardized GMT curriculum. To promote uniform acceptance and standardization (and potentially reduce costs), training could be linked from the DoD-wide website to a service website. Service training management systems could then use the standardized training. A DoD-wide accessible site could provide downloadable, flexible training options for unit commanders, such as lectures and web-based (CBT) options. Compatibility issues among LMSs would need to be addressed for CBT instruction. Standardized GMT content would increase interoperability and training transferability.

DoD should consider adopting standardized CBT for GMT. Collaborating on new and improved CBT could be a significant way to engage stakeholders. The services have already taken steps to standardize GMT requirements and reduce the GMT burden, and continued progress could be made thorough increased collaboration and sharing of ideas. Services could share what lessons they learned. Service SMEs desire options for training delivery. The sharing of information and best practices by all in the use of CBT can help the services provide well-designed, service-generic CBT versions (whenever feasible) to meet GMT requirements. The development and adoption of this CBT could be an important way for DoD to perform its oversight role.

DoD should perform a gatekeeper role for future GMT requirements. There are many training demands for uniformed DoD personnel, and adding new requirements carries with it both opportunity and financial costs. If new GMT requirements are levied across DoD, the magnitude of GMT requirements can grow, forcing unit commanders to make additional decisions between mission training demands and GMT-mandated requirements. New training requirements that are additive to existing requirements have a big effect by increasing the time required to meet training demands and are multiplicative across the force. Even valid GMT requirements must be balanced with the need to conduct individual specialty and unit collective training. The time impact of new GMT demands must be kept in check and balanced to provide time for mission and specialty training. A single DoD sponsor for GMT should compare new with current training demands to achieve this balance and reduce mission-readiness opportunity costs. The goal of this process would be to assess training requirements, increase effectiveness and cost-efficiency, resolve issues, balance priorities, and gain training efficiencies.

The services have indicated that GMT is a burden and that it can take time away from mission accomplishment. Our research examined both GMT directed by DoD and above and the training topics over which the services have direct control. In this report, we identify the extensive topics that the services require. The services have much discretion in reducing or eliminating these training requirements. If the services determine that GMT is a burden, an intraservice examination of training requirements may indicate that some of these requirements could be eliminated, reduced, or standardized.

DoD must engage with and get buy-in from GMT stakeholders. The stakeholders include the service leaders who develop GMT policy and service AC and RC training SMEs. Meetings with them could further assess opportunities to share ideas and consider potential standardization and reduction options. Training SMEs should be continually engaged to illuminate training issues and challenges and to share ideas. Service training programs, although developed independently of one another, have many similarities. The engagement approach with SMEs should focus on highlighting GMT similarities and reduce training differences between services.

All existing and new GMT requirements need to be challenged to determine whether they are needed or whether they could be standardized and reduced. DoD should engage with training-topic sponsors to validate training requirements. For example, the DoD equal opportunity (EO) office is in charge of the DoD EO instruction that requires periodic EO training (Department of Defense Directive [DoDD] 1020.02). The EO office, along with other training-topic sponsors, needs to be engaged to consider required content, periodicity, and reporting requirements for EO training. Involving the key stakeholders will serve to evaluate (and challenge) the training requirements, build consensus for change, and initiate change.

DoD should issue a single DoD directive that lists all GMT requirements. GMT demands are many; the aggregate effect on uniformed service members in terms of time demands across the force is high, especially on the reservists; and guidance that directs that training be completed comes from multiple DoD directives. A single directive or instruction would list GMT topics and periodicities and provide clarity to the services as they evaluate and further develop GMT programs.

Although the focus of this research was on GMT (done by all uniformed personnel), there is a great deal of training that is required by the services that could potentially best be performed by means of CBT for effectiveness and efficiency.

DoD, in conjunction with the services, should review the options that the services have implemented in their GMT curricula. Although the options or approaches taken appear to be having the desired effects of reducing the burden and standardizing requirements, they have not been analyzed to determine whether they have actually achieved these effects. Moreover, it has not been determined whether the options have undesirable consequences. Further analysis may suggest alternative or additional options to implement.

The Way Forward

The services have made efforts to standardize and reduce GMT. If OSD desires further standardization and reduction of GMT, it should lead this effort. To realize this opportunity, OSD should engage GMT stakeholders in a collaborative fashion to promote the sharing of successes and lessons learned. OSD might stimulate that collaboration by sponsoring a meeting in which the SMEs, responsible for the advances discussed in this report, discuss how they achieved those advances and lessons learned that would benefit other services.

Discussions or considerations about the reduction or standardization of GMT requirements should include an examination of the effectiveness and quality of the training. Training quality must be measured, understood, and maintained at a high level. Further research is needed to examine the quality and effectiveness of GMT. This additional study should evaluate the efficacy of different training delivery methods in terms of how well GMT information is transferred, i.e., what type of information or GMT topics best suit different delivery methods (i.e., stand-up instruction, CBT, or other), and delivery methods that are most effective for the intended audience. Such a study should evaluate the quality of the content and instructional design to identify additional opportunities to provide standardized training that can be prepared and delivered in less time than is required for current training programs.

Finally, new training requirements can pose significant time demands across the force. An OSD Office of Primary Responsibility (OPR) should be designated to facilitate information sharing that will help the services provide well-designed, service-generic training of all new requirements, whenever feasible. This could be an important way in which DoD performs its gatekeeper role.

Acknowledgments

We thank CAPT Robert Bennett, Deputy Director for Training Transformation, Office of the Assistant Secretary of Defense for Reserve Affairs, for his guidance and support during the course of the research. COL Jaime Lugo and Col Maureen Tritle also shared their time, insights, and suggestions during the research. We also appreciate the insights and assistance of CAPT Eric Young. We also thank COL Gary Kayser for his support in the final phase of the project. We gratefully acknowledge the assistance of the service representatives who provided valuable details of their respective ancillary and general military training programs. In particular, we acknowledge LTC Randal L. Mock, Baker Turner, Ladonna McGrew, and Maj Ryan Prince. We also appreciate the support provided by Marion Cain of the Office of the Secretary of Defense for Readiness and Training. We gratefully acknowledge the guidance and direction provided by Robert Smiley of the Office of the Assistant Secretary of Defense for Reserve Affairs.

This report benefited from the insightful reviews of colleagues John A. Ausink and Michael G. Shanley, who provided useful comments that helped strengthen the overall report. We appreciate the administrative assistance provided by Alvie Vickers.

Abbreviations

AC	Active Component
ACRM	advanced combat rifle marksmanship
ADLS	Advance Distributed Learning System
AFI	Air Force instruction
AFLC	Air Force Learning Center
AFR	Air Force Reserve
AFSC	Air Force specialty code
AFSO21	Air Force Smart Operations for the 21st Century
AIDS	acquired immunodeficiency syndrome
ANG	Air National Guard
APFT	Army Physical Fitness Test
AR	Army regulation
ARNG	Army National Guard
ASD RA	Office of the Assistant Secretary of Defense for Reserve Affairs
AT	antiterrorism
BAS	basic allowance for subsistence
BCRM	basic combat rifle marksmanship
BEPO	Base Emergency Preparedness Orientation
CBR	chemical, biological, and radiological
CBRN	chemical, biological, radiological, and nuclear
CBT	computer-based training
CCTV	closed-circuit television
CGBI	Coast Guard Business Intelligence

CHAMPUS	Civilian Health and Medical Program of the Uniformed Services
CI	counterintelligence
CMT	common military training
CoC	Code of Conduct
COCOM	combatant command
CONUS	continental United States
CPPD	Center for Personal and Professional Development
CTIP	combating trafficking in persons
DA PAM	U.S. Department of the Army pamphlet
DADLAT	Defense Advanced Distributed Learning Action Team
DCS	deputy chief of staff
DEERS	Defense Enrollment Eligibility Reporting System
DHS	U.S. Department of Homeland Security
DISA	Defense Information Systems Agency
DoD	U.S. Department of Defense
DoDD	Department of Defense directive
DoDI	Department of Defense instruction
DoDM	Department of Defense manual
EO	equal opportunity
EST	expeditionary skill training
FHA	Federal Housing Administration
FLTMPS	Fleet Training Management and Planning System
FM	field manual
FOIA	Freedom of Information Act
FY	fiscal year
G-3/5/7	Operations, Plans, and Training
GMT	general military training
HIV	human immunodeficiency virus
HQ USAF/A1	Headquarters U.S. Air Force Manpower and Personnel
IA	information assurance

ICRM	intermediate combat rifle marksmanship
IMI	interactive multimedia instruction
ISS	information security systems
IT	information technology
ITRO	Interservice Training Review Organization
J-7	Joint Staff Directorate for Joint Force Development
LMS	learning management system
MBST	Marine Battle Skills Training
MCBUL	Marine Corps bulletin
MCO	Marine Corps order
MCRP	Marine Corps reference publication
MCTIMS	Marine Corps Training Information Management System
MEO	military equal opportunity
MOS	military occupational specialty
MT	mandatory training
MWR	morale, welfare, and recreation
NATO	North Atlantic Treaty Organization
NBC	nuclear, biological, and chemical
NCO	noncommissioned officer
NDRI	RAND National Defense Research Institute
NKO	Navy Knowledge Online
NSC-NSDD	National Security Council national security decision directive
OASD RA	Office of the Assistant Secretary of Defense for Reserve Affairs
OCONUS	outside the continental United States
OMB	Office of Management and Budget
OPNAVINST	Office of the Chief of Naval Operations instruction
OPR	Office of Primary Responsibility
OPSEC	operations security
OSD	Office of the Secretary of Defense
OUSD(P&R)	Office of the Under Secretary of Defense for Personnel and Readiness

PCS	permanent change of station
PII	personally identifiable information
PR	personnel recovery
PRT	physical readiness test
RC	Reserve Component
SAEDA	Subversion and Espionage Directed Against the U.S. Army
SAFE	substance abuse–free environment
SAPR	sexual assault prevention and response
SAVI	sexual assault victim intervention
SECNAV	Secretary of the Navy instruction
SERE	survival, evasion, resistance, and escape
SETA	security education and training awareness
SGLI	Servicemembers' and Veterans' Group Life Insurance
SHAPR	sexual harassment and assault prevention
SHP	sexual harassment and prevention
SME	subject-matter expert
SOH	safety and occupational health
TFAT	Total Force Awareness Training
TIP	trafficking in persons
TMT	Training Management Tool
UCMJ	Uniform Code of Military Justice
VA	U.S. Department of Veterans Affairs

CHAPTER ONE

Introduction

Background

The purpose of this study was to establish a baseline of existing legislation, U.S. Department of Defense (DoD) and service guidance, and directives and policies that establish ancillary training and general military training (GMT) (herein, ancillary training and GMT are referred to as simply *GMT*) requirements for members of the armed forces (both Active Component [AC] and Reserve Component [RC]) and examine options to reduce and standardize these requirements. In this report, we analyze the services' current training approaches to meeting GMT needs, assess the GMT conducted throughout the services, and identify and evaluate opportunities to reduce and standardize training to meet GMT needs, with particular emphasis on reducing impacts on the RC pre- and postmobilization training burden in order to maximize available training time.

This report provides an overview of the services' GMT programs and identifies ancillary training and GMT requirements as directed by existing legislation, as well as by DoD policies, regulations, and instructions. GMT is a challenge because it is time consuming, all uniformed personnel must complete it, and it diverts time away from mission demands. DoD is interested in options to standardize and reduce mandatory requirements for both the RC and the AC. Standardizing and reducing GMT could provide common instruction (which would support training transferability) and save resources (by eliminating the need for redundant curriculum development). Standardization can also make meeting requirements more efficient and thus reduce resources and time required to do the training. Reciprocity or training transferability to another command because of standardization would reduce redundancy of training demands. Reducing GMT could minimize both the pre- and postmobilization training burden for the RC, which would increase the time that units could devote to mission demands.

Each service develops its own GMT programs. However, all uniformed personnel must complete some common topics. Although some topics are common among the service's GMT programs, many topics are unique to each service. The services vary in the number of GMT topics required. This report focuses on GMT that all uniformed personnel across DoD must complete.

The services also differ in how they deliver GMT. The services have the discretion to prescribe a preferred delivery method. Delivery methods range from traditional, instructor-provided training in a classroom to computer-based training (CBT). The preferred method of providing the training varies by service and, in some cases, by component within each service. For example, although one component prescribes traditional, instructor-led training, other components may prescribe other delivery options, such as CBT, to meet the various needs of the trainees.

Although most personnel understand the need for GMT, the effect of GMT on the RC's time is much less well understood. This report should assist in rectifying this deficiency and explore how the services differ in their approaches to GMT topics and how they deliver GMT.

Objective and Approach

The objective of the research was to define and determine the nature and scope of GMT currently being performed by the services' AC and RC and to identify options to standardize and reduce common cyclic mandatory training (MT) requirements. The purpose of identifying options to standardize and reduce GMT was to optimize available training time while meeting training demands. Note that this research did not evaluate standardization per se; it simply looked at ways in which training might be standardized. It did not explore the issue of whether standardization was a desirable approach or examine any potential drawbacks of standardizing training across the military services (e.g., loss of flexibility).

To meet these ends, researchers from the RAND National Defense Research Institute (NDRI) developed a working definition of *GMT*, investigated current training approaches used by the services to meet GMT needs, compared the common training conducted across all services, and evaluated and determined alternative opportunities to complete GMT requirements.

NDRI researchers also determined where training requirements originate—from higher authority (DoD or above) or from service or component. The research team identified directed (DoD, Army, Navy, Marine Corps, and Air Force) GMT requirements, including those stipulated by legislation, policies, regulations, and instructions. We also examined delivery methods prescribed by the services and assessed the pros and cons of these methods.

In addition, qualitative methods were used (e.g., interviews) to conduct the research. RAND analysts conducted a series of 11 interviews with services' AC and RC GMT training officials and subject-matter experts (SMEs) and participated in a facilitated discussion with the SMEs and the Office of the Secretary of Defense (OSD) sponsor. RAND analysts also met with the Defense Advanced Distributed Learning Action Team (DADLAT), Joint Staff Directorate for Joint Force Development (J-7) training representatives, and senior DoD officials.

The research involved several steps. The first step was to provide a working definition of *GMT* that could be used to categorize the services' training demands. The next step was to identify DoD- and higher authority (e.g., U.S. Code)–directed GMT and service-directed GMT. This research included an examination of legislation, policies, regulations, instructions, and delivery methods. The third step was to establish a baseline for the GMT requirements across the services, i.e., GMT that must be completed by all uniformed personnel. The fourth step was to assess the training requirements and delivery methods. We next identified strategies to standardize and reduce GMT and, finally, provide findings and recommendations.

NDRI researchers examined the nature and content of GMT performed by the services and assessed what constitutes GMT within each individual service and component. The research team proposed a working definition of *GMT* for use in examining service programs in follow-on steps. The research sponsor, the Office of the Assistant Secretary of Defense for Reserve Affairs (OASD RA), approved the researchers' proposed working definition of *GMT*, and we used the definition for completing the study tasks that followed.

RAND analysts examined and documented the delivery methods for completing GMT requirements in use or available for use within DoD and the services. The documentation of directed (DoD, Army, Navy, Marine Corps, and Air Force) GMT requirements, including those topics required by existing legislation, policies, regulations, instructions, and delivery methods, was provided to the sponsor as a project memorandum (Yardley, Woods, and Ip, 2009).[1]

The research included in-depth discussions with service SMEs, whose valuable feedback regarding service policies is included in this report. RAND researchers met separately with SMEs from the Army, Army National Guard, Army Reserve, Navy, Navy Reserve, Air Force, Air National Guard (ANG), Air Force Reserve (AFR), Marine Corps, and Marine Corps Reserve. We also met with Coast Guard and Coast Guard Reserve training officials to discuss the unique approach used by the Coast Guard for completing GMT requirements. The interviews consisted of a series of open-ended questions and discussion regarding the respective services' and components' challenges with, issues with, and approaches to completing GMT. SMEs discussed preferences for delivery of training and provided service reference materials.

The services developed GMT programs mostly independently of each other. In each service, efforts are under way to reduce time devoted to GMT. Although one training method might be considered an effective and efficient delivery method for one service (e.g., CBT for the Air Force), the view is not universal; other services prefer and prescribe another method (e.g., instructor-led training).

We established a baseline of common GMT topics across all services and identified training requirements that are similar. This document includes a matrix that links the services' training requirements that are similar or nearly identical to the applicable legislation, policies, directives, instructions, and regulations that apply to them, the periodicity, and methods of delivery. This list of training requirements is presented for consideration as the recommended baseline for DoD GMT subjects.

NDRI researchers assessed the training time expended by each service on GMT requirements and the effect of this MT time on the AC compared with that on the RC. The report includes a discussion of time that could be saved and benefits gained by adopting proposed reductions in GMT requirements.

Organization of the Report

In Chapter Two, we define *GMT* and provide a list of topics that meet the OASD RA–approved working definition of *GMT.* Chapter Three summarizes the approaches of the Army, Navy, Air Force, Marine Corps, and their respective components to completing GMT and provides a summary of meetings with service GMT SMEs. Chapter Four discusses why GMT is considered a challenge for the services. In Chapter Five, we list options to standardize and reduce GMT demands. Chapter Six provides our concluding observations. Appendix A contains our assessment of topics that meet the approved definition of *GMT,* Appendix B summarizes an OSD-hosted meeting with service GMT SMEs, and Appendixes C through F contain synopses of Army, Navy, Air Force, Marine Corps, and Coast Guard GMT programs, respectively.

---

[1] This document is not available to the general public.

CHAPTER TWO

Defining *General Military Training* and Establishing General Military Training Baseline Topics

This chapter both defines *GMT* and details the training conducted by the services that meets that definition. It also sets the context for GMT as it relates to other military training.

Defining *General Military Training*

We approached the first task, which was to define *GMT*, by conducting a literature review of the services' definitions and through discussions with the sponsor. The services do not use a common definition of *GMT*, and a DoD-wide definition does not exist. We examined the services' definitions of *GMT* to develop a working definition that the sponsor would approve.

We reviewed DoD's and the services' references to determine how *GMT* is currently defined. The term *GMT* is not universally used throughout DoD. Table 2.1 lists the terms and definitions used by DoD and the services.

Table 2.1
Training Terminology

Definition Owner	Term	Definition
DoD	Common training	Training that is not unique to a particular DoD component; training that has no special distinction or quality to an individual DoD component and is widely required (Department of Defense Directive [DoDD] 1322.18)
Army	CMT	Senior leader–selected general training requirements considered essential to individual or unit readiness. This training is common to all members of units, regardless of branch, career field, rank, or grade, and is incorporated into unit training programs as specified in the governing Army regulation (AR) (AR 350-1, Appendix G).
Navy	GMT	Nonoccupational general training required periodically for all Navy personnel and taught at the command level (Office of the Chief of Naval Operations Instruction [OPNAVINST] 1500.22E).
Air Force	Ancillary training	Universal training, guidance, or instruction, regardless of Air Force specialty code (AFSC), that contributes to mission accomplishment. It does not include functional, occupational, or additional duty training (Air Force Instruction [AFI] 36 2201, Vol. 1).
Marine Corps	Ancillary training	Training that supports individual readiness, such as suicide prevention, standards of conduct, voluntary education, and equal opportunity (EO) training. Ancillary training is directive in nature but should be prioritized below mission-oriented and formal training requirements when planning (Marine Corps Order [MCO] 1553.3A).

NOTE: CMT = common military training.

RAND analysts proposed a working definition, based on these definitions, of GMT to OASD RA, as follows:

> Ancillary training or GMT is periodic, nonoccupational directed training that provides common knowledge and skills required for all uniformed personnel. It enhances an individual's ability to (be prepared to) perform military duties or activities. An example of ancillary training or GMT is sexual assault awareness and prevention.

This working definition was derived from an examination of the services' definitions and includes elements of each. Periodic or recurring training is an element of all the services' GMT programs. *Nonoccupational* separates this training from those related to a specialty—i.e., it is training required for all. *Directed training* refers to training that was directed by an authority above the services (e.g., DoD-directed training). Training that provides common knowledge and skills to all uniformed personnel is an essential element of services' definitions. GMT enhances an individual's ability to perform military duties or activities related to an individual's readiness and contribute to mission accomplishment.

With an approved working definition of *GMT* established, we could then evaluate the services' GMT topics to determine those that met the approved definition.

General Military Training Is Common Training at the Most Basic Level

In broad terms, GMT can be thought of as providing foundational training for *all* service members. Each service has its own training programs, including specialty, equipment, and watch-standing training for uniformed personnel.[1] Figure 2.1 illustrates that GMT lies at the intersection of all service individual training requirements. GMT is common training at the most basic level, and the instruction provides knowledge that all uniformed personnel must have. The topics can be broadly categorized as providing training in the following areas:

- core values: Code of Conduct (CoC), EO
- institutional knowledge: combating trafficking in persons (CTIP)
- conduct required
  - how to take care of oneself: physical fitness, suicide prevention, substance abuse
  - how to treat others: sexual harassment, sexual assault prevention and response (SAPR)
  - guidance on security and safeguards: counterintelligence (CI) awareness, antiterrorism (AT), information assurance (IA), North Atlantic Treaty Organization (NATO) security, operations security (OPSEC), personally identifiable information (PII).

[1] *Watch-standing* refers to an assignment during which an individual has specific, detailed responsibilities on a recurring basis around the clock. An example of watch-standing is sentry duty.

Figure 2.1
The Relationship Between Service Training and General Military Training

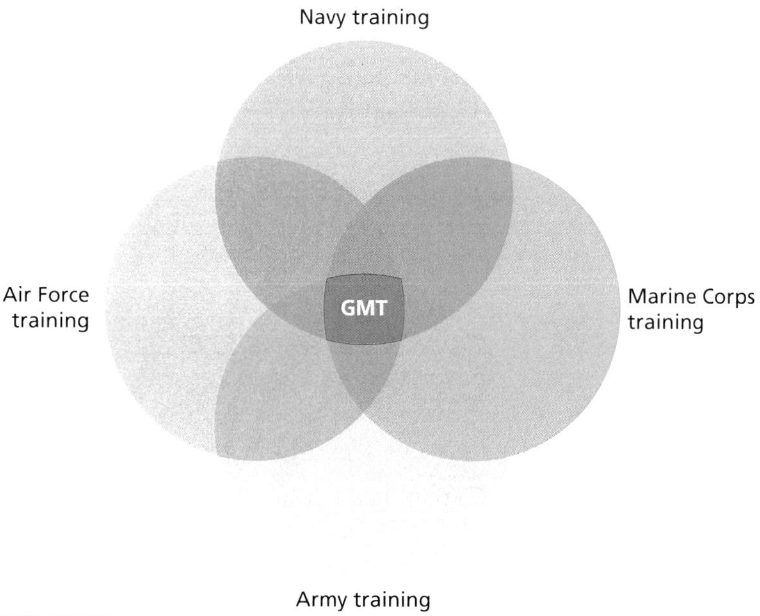

RAND *TR1222-2.1*

Identifying Training That Meets the Definition

After OASD RA approved the working definition of *GMT*, we examined each of the services' GMT programs to determine the topics that met the definition. The goal of this effort was to identify the GMT topics that are (or should be) common across all services.

We listed and examined every topic in service directives or given to us by service SMEs.[2] Appendix A lists Army, Navy, Air Force, Marine Corps, and Coast Guard GMT topics in alphabetical order and indicates the service that provides GMT on the topic. This matrix presents our assessment of each of the services' topics against each element of the working definition. This assessment was used to determine, by means of a yes-no decision, whether the topic meets the following six GMT definitional criteria:

- Is the training periodic?
- Is the training nonoccupational?
- Is the topic directed by DoD or above?
- Does it provide a common skill or knowledge?
- Is it required for all uniformed personnel?
- Does the training enhance the trainee's ability to perform military duties or activities?

[2] Although the Army, Air Force, and Marine Corps list GMT requirements in a standing instruction, the Navy does not. The Navy's GMT topics change from year to year, and Navy authorities transmit an annual GMT message directing required GMT for the current year.

For example, in Appendix A, the first topic is "alcohol misuse enabling." This was a Navy training topic, and our research indicated that it is not required training for all sailors to do periodically, is nonoccupational, is directed by DoD or above, provides common knowledge and skill, is required for all uniformed personnel, and enhances the trainee's ability to perform military duties or activities. We reviewed the content of the topic and applied judgment in including topics that, in our assessment, met DoD guidance. This topic fell under DoD guidance for substance abuse training, and we include it as a required training topic. Although this topic did not meet the periodicity criterion in accordance with Navy guidance (it is not required to be done periodically), our judgment was that it fell under substance abuse training and that, therefore, it is considered to be a GMT topic. We evaluated 132 topics in this manner.

The number of GMT topics is not uniform across the services. Moreover, the names of GMT topics are not standardized. We also found that, although a topic may be considered GMT for one service, it may not be considered GMT for another service. The training topics are developed independently based both on the needs of the service and per higher-level guidance. We recognize that each service develops training needs based on the demands of that service. Our approach was to evaluate how well the content of each of the services' topics met the criteria.

Some ambiguity surrounded our assessment regarding what topics to include as required training. We compared the guidance that directs the training with the elements of the working definition. In some cases, the DoD guidance was vague about training periodicity. For example, the DoD directives for suicide prevention and substance abuse training did not say how often training in those topics should occur. Even though the periodicity was not provided, we made judgments and discussed these topics with our sponsor. The judgments made and direction provided by the sponsor resulted in the inclusion of the following topics as meeting the criteria of the working definition:[3]

- AT
- CI awareness
- CoC
- CTIP
- EO
- IA
- NATO security
- OPSEC
- physical fitness
- Privacy Act (5 U.S.C. § 552)
- SAPR
- sexual harassment
- substance abuse
- suicide prevention.

---

[3] At the start of our effort, homosexual-conduct training was also required. During the study, the Don't Ask, Don't Tell policy (10 U.S.C. § 654) was repealed, so we went from 15 to 14 topics.

In addition, other training topics that were directed by DoD or above but did not meet all criteria of the working definition were chemical, biological, and radiological (CBR) training and training in safety and occupational health (SOH). The services are required to "routinely train all military personnel . . . (and others), in safety, first aid, sanitation, health risks, and health protection measures, including those related to chemical, biological, radiological, nuclear, explosive, and environmental and/or industrial threats" (DoDD 6200.04). Additionally, guidance on SOH training states,

> Military and DoD civilian officials at each management level shall advocate a strong Safety and Occupational Health (SOH) program; provide their personnel safe and healthful working conditions; and provide education and training that will enable them to prevent accidents, injuries, and occupational illnesses. (DoD Instruction [DoDI] 6055.1)

We again made a judgment. CBR and SOH training may be occupationally related or may need to be tailored to specific audiences, so they do not meet the definitional elements. Our research focused on the GMT topics that meet the criteria of the GMT definition. Although we recognize that other GMT is done by the services, our focus was to identify options to standardize and reduce GMT topics that meet the definitional criteria of GMT. As such, the remainder of our analysis focuses on those topics judged to meet the definition.

In this report, we briefly consider the magnitude and burden of other topics, and many of the recommendations presented for the primary topics on which we focused apply to the other topics. The standardization and reduction considerations for these topics are the domain of the services. That is, although GMT topics are directed from authorities above the service, the services have flexibility and discretion in the standardization and reduction of other GMT topics.

References and guidance that direct GMT were examined to understand the goal of the training, by topic. Table 2.2 lists these training topics and details the aim or goal of the training provided in the guidance.

Table 2.3 lists these topics and the DoD or higher guidance that directs the training. The table lists the broad guidance contained in the directive and the periodicity with which the training must occur, where given.

Table 2.4 lists the GMT periodicity per the DoD guidance. The training periodicity is explicit (e.g., annually) for some topics and less explicit (e.g., throughout one's career) for others. A lack of explicit guidance leaves the periodicity open to interpretation and provides the services flexibility and variability. In some cases, more-explicit guidance may be needed regarding training periodicity. For example, the periodicity of suicide prevention is not provided in the guidance, yet it is a topic of high concern to DoD. Even with explicit guidance, some GMT topics are not included in services' documented GMT plans. More guidance in the form of a common DoD directive could reduce uncertainty and provide clear direction to the services about when (and where) training is required (e.g., annually or at specific points during one's career).

Summary

The developed, working definition of *GMT,* approved by OASD RA, was established to examine the services' training programs and to determine topics that met the definition of *GMT.*

Table 2.2
General Military Training Topics and Aim of Training

Topic	Aim of Training
AT	Provides basic techniques and strategies to avoid becoming an easy target; teaches proven survival methods
CI	Promote threat and reporting awareness responsibility; enable DoD personnel to identify CI threats and report suspicious situations and incidents to appropriate authorities
CoC	Foster motivation and dedication necessary for service members to survive captivity and hostile action; promote compliance with the law of war and preserve the integrity of the Armed Forces by reminding service members of the highest values the military services embody, including love of the United States and faith in its democratic ideals and institutions; derived from Executive Order 10631
CTIP	Educate all on the worldwide trafficking menace, national TIP policy, overseas theater TIP policy, and attendant personal responsibilities consistent with DoD core values and ethical standards
EO	Training focuses on the right of all to participate in and benefit from programs and activities for which they are qualified, free from social, personal, or institutional barriers that prevent people from rising to the highest level of responsibility possible
IA	Provides initial IA orientation and annual awareness training to all authorized users to ensure that they know, understand, and can apply the IA requirements of their systems
NATO security	Reinforce the policies, principles, and procedures covered in initial and specialized training. Refresher training should also address the threat and the techniques employed by foreign intelligence activities attempting to obtain classified information and advise personnel of penalties for engaging in espionage activities.
OPSEC	Reinforce OPSEC policies and procedures, critical information, and procedures covered in initial and specialized training. Refresher training addresses the threat and techniques employed by adversaries attempting to obtain classified and sensitive information.
Physical fitness	Stresses the benefits of appropriate physical activity and nutrition in maintaining and enhancing health, physical fitness, safety, and injury prevention
Privacy Act/PII	Provides information regarding privacy laws, regulations, policies, and procedures governing DoD's collection, maintenance, use, or dissemination of personal information. The objective is to establish a culture of sensitivity to, and knowledge about, privacy issues involving individuals throughout DoD.
SAPR	Training geared to prevent and eliminate sexual assault by providing comprehensive procedures to better establish a culture of prevention, response, and accountability that enhances the safety and well being of all DoD members
Sexual harassment	Training on policies and programs on the prevention of sexual harassment
Substance abuse	Promote and improve the health of the force through the prevention of substance abuse
Suicide prevention	Provide knowledge of resources to increase awareness of suicide concerns, improve wellness, and clarify how to intervene when someone needs help; ensuring timely access to needed services and having a plan of action for crisis response; assisting families, units, and sailors affected by suicide behaviors; and report incidents of suicide and suicide-related behaviors

NOTE: TIP = trafficking in persons.

We examined each of the services' GMT topics and weighed them against criteria developed to determine whether the topic met the OASD RA–approved working definition of *GMT.* Refer-

Table 2.3
General Military Training Topics, Guidance That Directs Completion, and Periodicity Requirements

Topic	What the Guidance States
AT	"The Heads of the DoD Components shall: Provide post accession Level I AT Awareness Training annually to all DoD personnel." (DoDI 2000.16)
CI awareness	"It is DoD policy that: The DoD personnel shall receive periodic briefings on the threats posed by foreign intelligence services, international terrorists, computer intruders and unauthorized disclosures, and individual reporting responsibilities. This shall include insider threats and the crimes of spying and treason. Briefings shall be presented at or near the time of initial entry or hire and thereafter at least every 12 months. More frequent briefing intervals should be instituted if conditions warrant. Some DoD Component organizations or personnel may require more frequent briefings predicated on the nature of their duties." (DoDI 5240.6)
CoC	"Indoctrination in the CoC shall begin without delay on the entry of Service members into the Armed Forces and shall continue throughout their military careers. . . . The Military Services are responsible for adapting a comprehensive CoC training program to support the requirement of the COCOMs. . . ." Executive Order 10631 states that "members of the armed forces liable to capture shall be provided with specific training and instruction designed to better equip them to counter and withstand all enemy efforts against them, and shall be fully instructed as to the behavior and obligations expected of them during combat or captivity." (DoDD 1300.7)
CTIP	"Educate all Service members and DoD civilians annually on the worldwide trafficking menace, national TIP policy, overseas theater TIP policy, and attendant personal responsibilities consistent with DoD core values and ethical standards." (DoDI 2200.01)
EO	The DoD MEO program shall "[p]rovide periodic, mandatory education and training in human relations and MEO at installation and operational unit commands, during pre-commissioning programs and initial entry training, and throughout professional military education systems as part of the overall effort to achieve MEO within the Department of Defense." (DoDD 1020.02)
IA	"Provide for initial IA orientation and annual awareness training to all authorized users to ensure they know, understand and can apply the IA requirements of their system(s). . . ." (DoDD 8570.01-M)
NATO security	"Cleared members . . . and members of the organization who are not cleared for access to classified information should be included in the security education program if they will be working in situations where inadvertent access to classified information might occur or will have access to unclassified information that might be of value to intelligence collectors." As a minimum, personnel are to receive annual refresher training that reinforces the policies, principles, and procedures covered in initial and specialized training. (DoDD 5200.1-R)
OPSEC	"OPSEC education should be continuous rather than periodic. . . . As a minimum, all personnel shall receive annual refresher OPSEC training that reinforces understanding of OPSEC policies and procedures, critical information, and procedures covered in initial and specialized training." (DoDM 5205.02-M)
Physical fitness	"All Service members, regardless of age, will be formally evaluated and tested for record at least annually unless under medical waiver." (DoDI 1308.3)
Privacy Act/ PII	"Components shall give consideration to whether annual training and/or annual certification should be mandated for all or specified personnel whose duties and responsibilities require daily interaction with personally identifiable information." (DoDD 5400.11R)
SAPR	"The Military Services shall: Provide periodic, mandatory education in SAPR at installation and fleet unit commands. . . . Conduct an annual sexual assault awareness training update. To help Service members understand the nature of sexual assaults in the military environment, this training should be scenario-based, using real-life situations to demonstrate the entire cycle of reporting, response, and accountability procedures." (DoDI 6495.02)

Table 2.3—Continued

Topic	What the Guidance States
Sexual harassment	"The Secretaries of the Military Departments are responsible for EO and for ensuring compliance with this Directive within their respective Services and shall: Ensure that all military personnel, including command selectees and flag and general officers, receive adequate training in EO, human relations, and prevention of sexual harassment on a recurring basis, and at all levels of PME." (DoDD 1350.2)
Substance abuse	"Promote and improve the health of the force through programs on injury prevention, dental health, good nutrition, tobacco use prevention and cessation, physical fitness and weight control, responsible sexual behavior, stress management, suicide prevention, alcohol and drug abuse prevention, and other health initiatives" (no periodicity given). (DoDI 6200.04)
Suicide prevention	"Promote and improve the health of the force through programs on injury prevention, dental health, good nutrition, tobacco use prevention and cessation, physical fitness and weight control, responsible sexual behavior, stress management, suicide prevention, alcohol and drug abuse prevention, and other health initiatives" (no periodicity given). (DoDI 6200.04)

NOTE: Red text indicates periodicity. DoDM = DoD manual. COCOM = combatant command. MEO = military EO.

Table 2.4
General Military Training Topics, by Periodicity Requirement

Periodicity	Topic
Annual	AT, IA, OPSEC, NATO security (cleared members), CTIP, CI awareness, SAPR, physical fitness
Throughout career; periodic; recurring basis; give consideration to annual training	CoC, EO, sexual harassment, Privacy Act, and PII
No periodicity	Suicide prevention, substance abuse

ences were examined to determine the guidance that directs that a topic be completed and how often. Fourteen GMT topics were judged to meet the definition of *GMT*.[4]

Training periodicity contained in DoD guidance is ambiguous for some topics. The services have discretion in determining how service members or others complete the service-directed GMT. Our focus is to identify options to standardize requirements for and reduce the burden of GMT for topics that meet the definitional criteria of *GMT*.

In the next chapter, we examine and summarize the services' approaches to GMT programs.

---

[4] At the start of our effort, homosexual-conduct training was also required. During the study, the Don't Ask, Don't Tell policy was repealed, so we went from 15 to 14 topics.

CHAPTER THREE

Summary of Service Approaches to General Military Training

After categorizing GMT and determining which topics were most directly under the purview of the services, we examined each of the services' GMT programs and interviewed service training SMEs to understand their objectives, training approaches, and how the training is delivered. The goal of this review was to gain a better understanding of the services' programs in order to identify potential options to standardize GMT requirements and reduce GMT burden. We examine and summarize the services' programs and approach and discuss our individual meetings with service SMEs. OSD sponsored a group meeting with the services' GMT SMEs to discuss potential options to standardize and reduce the GMT burden, and we provide a summary of this discussion. Appendix B provides further information about the discussion that occurred at the OSD-sponsored meeting with service SMEs. In Appendixes C through F, we also provide greater details on Army, Navy, Air Force, Marine Corps, and Coast Guard (including guard and reserve) programs, respectively.

Summary of Service General Military Training Programs

Army

The Army provides a great deal of GMT to its troops. The Army currently provides training in 21 GMT topics. The Army's preferred delivery method is instructor-led stand-up delivery by a senior noncommissioned officer (NCO) or officer. This training method reinforces the core values contained in the training topic because it affords Army leaders the opportunity to exercise leadership and mentorship (and conversations) with the troops. The number of topics on which the Army provides GMT is large. The time it takes to complete each training topic ranges from 30 to 60 minutes. Consequently, the Army spends a considerable amount of time on GMT. Training SMEs stated that much of the training is locally prepared and that training materials are not standardized across the Army.

Navy

The Navy recently made deep cuts in its GMT program. In the past, all uniformed personnel were required to be trained in ten to 11 topics annually. The Navy scaled GMT back significantly and currently requires all hands to receive annual training in only five GMT topic areas:

- responsible use of alcohol (use, abuse prevention, and control)
- responsible personal behavior (SAPR, EO, and sexual harassment grievance procedures)
- ask-care-treat (suicide awareness and prevention)

- improving personal financial management
- operational stress control.

The Navy-prescribed delivery method for these topics is instructor-led training sessions at each unit by leadership or command training teams composed of officers and senior enlisted personnel. Instructional materials for training instructors Navy-wide are downloaded from the Navy's learning management site (the Navy Knowledge Online [NKO] website). These materials are maintained and updated as needed by the Navy.

The Navy removed three topics from the annual GMT (hazing policy and prevention, fraternization awareness and prevention, and homosexuality policy) but requires training in these topics upon reenlistment. The training materials are provided on the NKO website. Family readiness program training was added as a reenlistment training topic in fiscal year (FY) 2011 (NAVADMIN 098/10).

The Navy may have gone too far in reducing its GMT requirements. Some topics that are judged to be GMT requirements per the approved working definition of *GMT* are not included in the Navy's GMT guidance (NAVADMIN 098/10). Making further reductions in GMT is not a likely alternative for the Navy.

The Navy also has plans to restructure the training to specific audiences (nonsupervisory, supervisory, and command leadership).

Air Force

The Air Force calls GMT *ancillary training*. There are 29 Air Force training topics. The Air Force recognized the impact that ancillary training has on airmen's time and has made a dedicated effort to reduce ancillary training time demands. It has done so by dividing ancillary training into four categories:

- annual Total Force Awareness Training (TFAT): general awareness training for the total force, delivered through the Air Force's LMS, Advanced Distributed Learning System (ADLS), at 12-month intervals
- selected force training: training to specific audiences, such as commanders and supervisors
- event-driven training: training keyed to a specific event, such as reporting to a new unit
- expeditionary skill training (EST): training to prepare an airman to serve in a contingency environment.

The Air Force minimized the amount of TFAT required of all personnel, tailored other training to be given only to selected audiences, made some training event-driven (rather than annual), and provided EST to those preparing to serve in operational environments. The Air Force approach to reducing and tailoring ancillary training could potentially be a model for other services.

The Air Force–preferred method of training (for most topics) is online, but some topics must be instructor led (e.g., SAPR). Air Force RC SMEs indicated that instructor-led training could be problematic for reserve units. The challenges include finding available classroom space and finding qualified instructors to conduct the training on drill weekends.

Marine Corps

The Marine Corps is closely examining its GMT program. The Marine Corps, like the Army, prefers and prescribes the instructor-led stand-up delivery method because it provides Marine Corps leaders the opportunity to exercise leadership and mentorship with the troops and to reinforce core values. The Marine Corps has adopted some Army CMT as its own—specifically, OPSEC, CoC, and CI awareness. Because the Marine Corps operates in the same or similar environments to the Army, adopting Army training in these topics is a logical move (and saves time developing curriculum).

The Marine Corps spends a significant amount of time performing GMT. Currently, the number of Marine Corps GMT topics is 48. Unlike the other services' topics, a majority of Marine Corps GMT is service-directed. However, this does not necessarily translate into additional training time because not every topic is designed to be completed every year. We met with a Marine Corps SME responsible for evaluation and development of GMT policy. The discussion indicated that the Marine Corps is closely examining opportunities to reduce and standardize GMT topics. Currently, the number of Marine Corps GMT topics is 48. Unlike the other services' topics, a majority of Marine Corps GMT is service-directed. However, this does not necessarily translate into additional training time because not every topic is designed to be completed every year.

Coast Guard

The Coast Guard calls GMT *MT*. In peacetime, the Coast Guard falls under the U.S. Department of Homeland Security (DHS) rather than DoD. We include the Coast Guard here because it has an innovative approach to completing MT. All of the Coast Guard's seven MT topics are required by an authority higher than the Commandant of the Coast Guard.

The Coast Guard has put considerable effort into moving all its mandated training topics to its online LMS. Although training may be completed in classroom (i.e., stand-up) delivery, the preferred method is online, interactive CBT for all topics, which requires users to answer questions as they proceed through the training. All Coast Guard CBT is interactive, and all requires satisfactory completion of a test to complete the course.

The Coast Guard expects to redesign all its online courses to present a customized experience based on a pretest. If a satisfactory score is achieved on the pretest, the training topic is recorded as completed, and the user receives no further training on that topic. If an unsatisfactory score is achieved on a particular section or sections of the pretest, the user is directed to a customized version of the training designed to improve his or her knowledge on the specific topics for which the scores were unsatisfactory. If new content is added to an individual MT curriculum, the system can be tailored so that a user can either be required to take only that section of the training or be given the opportunity to test out of the training. For example, the Coast Guard expects the average time for a service member to complete all seven topics in the redesigned format to be four hours. However, one Coast Guard member with 15 years of experience completed all MT in less than two hours by "testing out" of much of the subject matter.

This innovative approach, using an interactive method, supports completion of training and eases tracking and reporting.

Summary of Individual Meetings with Service General Military Training Subject-Matter Experts

The services have independently developed GMT programs geared to provide MT to their personnel. To better understand services' approaches, we interviewed GMT SMEs, including representatives from these organizations:

- Army
- Army Reserve
- Army National Guard
- Navy
- Navy Reserve
- Air Force
- AFR
- ANG
- Marine Corps
- Marine Corps Reserve
- Coast Guard.

The interviews consisted of open-ended questions about service GMT programs. We asked the SMEs to relate their knowledge of service approaches to GMT, number of topics, content, delivery methods, challenges, and time needed to complete GMT. Service SMEs also provided us with reference materials that govern service GMT programs.

Service General Military Training Programs Are Tailored to Each Service

Service GMT programs include both service-directed training and training directed by DoD (and above). The services develop their GMT independently of one another. Topics that are considered to be GMT for one service may not be a GMT topic for another service. Overall, we found that, even among common topics, a stove-piped approach is used in that GMT is planned, developed, conducted, and tracked independently by each service.[1] As indicated in the previous section, the number of GMT topics varies by service; however, some GMT topics are required for all uniformed personnel.

Although the number of topics varies by service, GMT topics—according to our working definition of *GMT*—are required to be done across the board by all uniformed service members. For some services (i.e., Navy and Air Force), the majority of the GMT topics are required by DoD or above. For others (i.e., Marine Corps), the majority of the GMT topics are required by the service.

The Reserve Component Follows Active Component Guidance

An overall observation is that the RC follows AC GMT guidance. In each service, one overarching directive indicates that GMT demands are the same for both the AC and RC.

---

[1] The services have revised the number of training requirements since Appendix A was developed. The Navy has greatly reduced the number of GMT topics.

General Military Training Is Not Universally Accepted

GMT completed at one command is not necessarily accepted at another. Several of our inter viewees recounted situations in which they completed training at their parent commands and, upon transferring to another command (within a service or to a joint command), they had to complete the same GMT again. The officers indicated that this was easier than pursuing the training officer to verify completion at their previous command. The officers expressed frustration and dissatisfaction with having to retake training that they had already completed because this consumes the trainee's valuable time.

Challenges Differ Between the Active and Reserve Components

The RC and AC face different challenges in completing required training. First and foremost, RC personnel have less time to complete the training because they are available only one weekend per month for drills and two weeks per year for training. Additionally, there is a lack of resources to do the training. RC (and some AC) experts stated that there are limited classrooms available to do required stand-up lectures; in addition, there are a limited number of computers available to do training that can be done only online using CBT. Finding time and the resources to do the training was an issue repeatedly raised during our research.

Both General Military Training Delivery Methods and Quality Vary

Preferred delivery options vary among the services, and SMEs voiced that they desire options when it comes to delivery methods. GMT topics can be taught through a variety of delivery methods. As mentioned earlier, the Army-, Navy- and Marine Corps–prescribed delivery method is instructor-provided training. The Air Force and Coast Guard use CBT as the primary GMT method.

When reviewing the GMT CBT topics, we observed a wide range of quality levels in instructional design. The quality of CBT is important because it has an impact on the amount of time required to teach the desired content. Increased training quality can potentially reduce training times. Although the measurement of the quality of training is complex, higher-quality training is thought to be training that is dynamic and interactive and has clear learning goals. It encourages the trainee to engage with the content (Codone, 2001). For the most part, much of the older training that we examined appeared to be of the lower-quality "click-through" design, while newer training appeared to be of higher quality. The older training provided very little interaction with the trainee. The more-recent and updated CBT appeared to be of higher quality and to have greater interaction with the trainee. If a pattern of higher-quality CBT emerges, then the amount of time that will be spent on GMT completed via CBT will naturally reduce as older courses are replaced with newer ones, and higher training quality may result.

Any efforts to standardize requirements and reduce GMT burden must be balanced with the quality of the training. The quality of GMT must be maintained at a high level to meet the training goals, and training quality must not be sacrificed at the expense of standardization and reduction efforts.

SMEs voiced that there are few measures of GMT quality or effectiveness. Although training comprehension can be gathered by interactive responses to a well designed GMT CBT course, measures for collecting those data are not in place.

Much training is given using stand-up delivery. In some cases (e.g., suicide prevention), the service requires that well-trained and presumably competent instructors conduct the training. In others (e.g., SAPR), trainers are required to hold training in small groups. But, for the most part, an officer or NCO presents the topics via the stand-up method to service members in his or her chain of command. There is, of course, the potential for a wide variety of quality levels for stand-up delivery, depending on the experience and training expertise of the trainer. Some training is also delivered via closed-circuit television (CCTV), e.g., onboard Navy ships. Qualified instructors or senior leadership can meet a wide audience at various locations using this method. The training can be recorded and shown again for personnel who missed the training (e.g., those on watch).

Producing common CBT may be an appropriate strategy for adding delivery options and gaining more GMT standardization. The services do have some CBT versions of GMT, and the Air Force and Coast Guard use CBT as the primary delivery method. The quality, content, and instructional design are approved for use by these services. CBT versions of GMT should be evaluated for potential wider use by all services.

Preferred Delivery Methods Vary

Overall, the Army, Navy and Marine Corps prefer stand-up instruction, i.e., instruction delivered by an individual or group of instructors. The Air Force and Coast Guard prefer CBT as the primary delivery method. Service SMEs say that the reason for this preference is that instructor-led stand-up instruction affords an opportunity for leadership and mentorship between instructor and student, i.e., between senior and junior NCOs or between officers and NCOs. This type of training reinforces core values and the connection between troops and leaders. However, instruction via CBT also has advantages, which are discussed later in this report.

Delivery Options for General Military Training Are Desired

GMT requirements are derived from the same DoD instruction or higher guidance, but the training delivery method is not standardized across services. Although CBT delivery may be optimal for one service, it may not be optimal for another. As noted, RC SMEs expressed a desire for delivery options. The components discussed delivery options for GMT during an OSD-sponsored GMT meeting.

Some services require face-to-face instruction for specific topics. For example, suicide prevention training, although available by means of CBT, is normally given by a health care provider in the Navy and similarly qualified personnel in other services. Additionally, the Air Force directs that SAPR be given in small groups by qualified SAPR instructors. This requirement is not uniform across all services.

Moreover, in one service (the Army), the training is locally developed. Although locally developed training may allow the training to be tailored to fit the situation in a particular geographic area, such a narrow focus may miss the broad objectives of training and can result in unstandardized training within a service.

The Services Have Websites for Online General Military Training

Each of the services has its own LMS through which service members can complete at least some online training

- Army officials noted that they are transitioning to the Army training network, the U.S. Army Distributed Learning System (see U.S. Army Distributed Learning System, undated). The Army still prefers stand-up delivery, however.
- The Navy uses NKO as a portal for downloading command presentations (instructor guides and materials) for standardized presentations, though it prefers stand-up delivery.
- The Air Force's online GMT is found on ADLS, and this is its preferred delivery method.
- The Marine Corps uses MarineNet for online training, though it prefers stand-up delivery.
- The Coast Guard requires its personnel to complete all GMT online, and its LMS, its sole method for delivering GMT, is the Coast Guard Learning Portal.

Summary of the Office of the Secretary of Defense–Sponsored Meeting with Service Subject-Matter Experts

At the OSD-sponsored meeting, the service GMT SMEs expressed a desire for tailorable training that could be provided to different audiences. They prefer to have options to provide training through various delivery methods, which would provide flexibility to unit leadership in meeting training demands. SMEs suggested that some similar topics could potentially be bundled. In addition, the SMEs suggested that some training be event-driven (e.g., upon reenlistment or permanent change of station [PCS]). The SMEs also discussed the issue of what the "right" periodicity was for conducting training. Although many training topics are listed as annual, the question posed was whether the annual training requirement could be extended (or reduced based on content or operations).

At the OSD-sponsored meeting, all participants said that they thought that a single DoD regulation listing all GMT would be helpful. Currently, all topics are listed in different DoD instructions and directives, and it is hard to keep track of all the requirements. Our research indicates that what the services consider directed training requirements are not consistent across the board. There was general agreement that it would be very helpful if one DoD instruction were issued that consolidated and standardized current GMT guidance.

Overall, GMT SMEs expressed the need for tailorable training, a variety of delivery methods, and flexibility for unit leadership in determining training delivery methods. The SMEs voiced the opinion that DoD should look to challenge every aspect of training levied on the force (e.g., is the training still needed, by whom, at what periodicity, and why). Moreover, the SMEs stated that options to reduce the GMT burden should be pursued by examining opportunities to bundle or combine some topics, determining whether training could be event-driven, and challenging the training periodicity.

Our Assessments of Delivery Methods and Varying Time Requirements

Delivery Methods

Our evaluation of the services' GMT programs included an examination of GMT delivery options used and an evaluation of the pros and cons of these options. The main delivery methods that the services choose are stand-up delivery and CBT.

Stand-up delivery is instruction in which an instructor or leader conducts a training session with a group of service members. The Army's, Navy's, and Marine Corps's preferred delivery method is the stand-up delivery method. The Navy provides some standardization of its stand-up material through its LMS. Its command presentation format, in which instructors use the LMS to download standardized instructional materials, typically provides slides to be used as talking points, and discussion sometimes follows the presentation. Attendance is sufficient to meet the training requirement, although a test, if available, can help measure training effectiveness.

CBT can be split into two categories: interactive and noninteractive. An example of interactive CBT is the Army's interactive multimedia instruction (IMI), which is interactive, electronically delivered training. IMI includes computer-controlled or computer-run courses that rely on student input to determine the pace, sequence, and content of instructional delivery. PowerPoint presentations that an individual reads and "clicks through" to receive training are considered noninteractive.

Both stand-up and CBT delivery methods have advantages and disadvantages, which we compare in Table 3.1. The services as a whole, especially the reserves, desire having multiple delivery options so that each unit commander can decide what type of delivery method is appropriate for the topic and his or her unit and how to use the available time most effectively.

The advantages of stand-up delivery are that face-to-face interaction can be supported and that participants have the opportunity to ask questions or engage in discussion. The services use this method to exercise leadership and mentorship, which allows leadership the opportunity to reinforce core values and ethos. Experienced instructors can gauge comprehension,

Table 3.1
Pros and Cons of General Military Training Stand-Up Delivery Versus Computer-Based Training

	Stand-Up Delivery	CBT
Pros	Face-to-face interaction can offer the opportunity to ask questions or engage in discussion. It provides the opportunity to exercise leadership and mentorship. It allows leadership to reinforce core values and ethos. The instructor or trainer can gauge comprehension and can test comprehension by tests. It offers a personal touch. It is customizable and updatable on the fly. Irrelevant or outdated training content can be removed easily. Locally and culturally relevant content can be included. Specialized facilities and equipment, such as a sufficient number of computers or computer labs, are not needed.	Training content can be standardized. Comprehension can be tested. Progress can be electronically tracked and reported. The flexibility provided means that training can be done anywhere at anytime and on a service member's own time. It offers the potential for pretest and test-out. It offers the potential for customized delivery, depending on the participant's pretest knowledge level. With sufficient existing infrastructure and investment, CBT can save resources—time and money.
Cons	Facilitators' skill levels can vary. Content is not standardized. It takes time to prepare and deliver instructional material. Large-group training settings can inhibit questions. Comprehension can be tested, but such testing is time-consuming. Muster sheet–based tracking Progress tracking is done at the unit level and is therefore decentralized. Reporting to higher levels of command can be unwieldy. It can be more expensive than CBT.	It can seem impersonal. Participants have no opportunity to ask immediate questions. If it is noninteractive, the participant can quickly click through training. Sustainment costs are incurred for updates. Units and individuals can have limited resources (computers). If done on a participant's own time, limited funds may be available to compensate individuals. It requires an existing network infrastructure with computers with sufficient availability to allow service members to access and complete training in a timely fashion.

and comprehension can also be gauged by tests. This method allows for a personal touch on the training topic and can be customized on the fly. Irrelevant or outdated training content can be easily removed, and the training can be tailored to include locally or culturally relevant content.

The disadvantages of the stand-up delivery method of training are that the facilitators who lead instruction may have varying levels of instructional skills, the content may not be standardized, instructional material is time-consuming both to prepare and to deliver, and training given in large-group sessions can inhibit questions. In addition, although comprehension can be tested, formal testing is time-consuming. Finally, attendance is tracked via muster sheets, which can be unwieldy.

The advantages of CBT are that it provides standardized training content; comprehension can be electronically tested, tracked, and reported; and the training can be done anywhere at anytime and on a service member's own time. With CBT, there is potential for pretest and test-out, as well as the potential for customized delivery, depending on pretest knowledge level. CBT has the potential to save resources—time and money. In general, review of the literature with respect to efficiency of CBT conservatively indicates a 30-percent reduction in training time when the same objectives are taught on computers in comparison with conventional instruction (Kulik, 1994; Sitzmann et al., 2005). CBT is efficient because personnel learn at different speeds. In instructor-led training, everyone proceeds at the same speed; with CBT, learners can work through training at their own pace. Increased retention can also result with CBT because the participant can spend the time needed on a topic that is challenging for him or her to comprehend.

The benefits of CBT will only accrue with high-quality training designs and sufficient infrastructure (computers) to use this mode. Stand-up delivery is used to exercise leadership and mentorship in communicating the "core values" covered in some important topics. However, if the services are seeking to reduce the overall GMT burden, well-designed, high-quality CBT is one option with the potential to both standardize the presentation and reduce time spent training on a particular topic.

The disadvantages of CBT are that it is impersonal and there is no opportunity to ask immediate questions. If the CBT is noninteractive, a trainee can quickly click through the material without necessarily comprehending it. Sustainment costs are incurred for updating the training material. Training may not be readily available for all personnel because units and individuals can have limited resources (lack of computers). If GMT were done via CBT on an individual's own time, i.e., for RC personnel, only limited funds would be available to compensate individuals for their effort. CBT also requires existing network infrastructure with computers with sufficient availability to allow service members to access and complete training in a timely fashion.

The Time Required to Complete General Military Training Varies

The services place different levels of emphasis on GMT topics; as a result, the time allotted to a topic and the mode of delivery vary across the services.

We examined service directives and LMS websites to determine the required training topics, the notional time needed to complete GMT per individual, and whether the topic could be taught using CBT. When service guidance did not list the time it takes to conduct the training, we consulted with the service SMEs.

Table 3.2 indicates the topics that fit the OSD-approved working definition of *GMT* and the amount of time (in hours), by service, that is estimated to complete the training. The topics that are part of the services' published GMT *and* available as CBT are highlighted in light blue. The shaded cells are topics that *are not* listed as service-required GMT but are available on a service's website through CBT. The unshaded (white) cells are topics not included in a service's GMT training plans (but may be addressed by the service elsewhere). The Army provided a range of time that may be needed to complete some training topics. Air Force CI training is required every four years. One hour of Air Force substance abuse prevention training is required once every four years; the Coast Guard requires 45 minutes of substance abuse prevention training every three years); the Army requires 30 to 60 minutes of EO training twice per year.

Table 3.2
Estimated Time, in Hours, to Complete General Military Training Topics

Topic	Army	Navy	Air Force	Marine Corps	Coast Guard
AT/FP	0.5–1		0.5	1	0.75
IA	2		1	2	
OPSEC	0.5–1	Recommended	0.5	0.5	
NATO security				0.5	
Privacy Act	0.5–1			2	0.75
CoC	0.5–1	Recommended		0.5	
CTIP	0.5–1		0.5	0.5	
Suicide prevention	0.5–1	1		4	0.75
Substance abuse prevention	0.5–1	1[a]	1 every 4 years		0.75 every 3 years
CI awareness	0.5–1		0.5	0.5	
EO	0.5–1 twice per year	1	1	1[b]	
Sexual harassment	0.5–1 twice per year				0.75
SAPR	0.5–1		1[b]	1	
Homosexual-conduct policy[c]	0.5–1	Upon reenlistment	1	b	
Physical fitness	Service specific				
Sum	9–16	3	6.25	13.5	3.25

NOTE: FP = force protection. Blue shading indicates a topic found on the CBT site and part of the service's published GMT. Gray shading indicates a topic available as CBT and not part of the service's published GMT. Cells spanning multiple rows indicate topics that the service has combined for training.

[a] Training on responsible use of alcohol is required. Training on drug abuse and tobacco use prevention is recommended.

[b] Service GMT topic, not available via CBT.

[c] Recent legislation repealed this policy.

The presence of shaded cells (not part of service GMT) or white cells (no topic listed in GMT guidance) may indicate a disconnect between our assessment of what the required GMT topics should be and what topics are included in service GMT plans. Table 3.2 shows a time only if both the service and DoD require the training. We note that the services (especially the Navy and Coast Guard) do not require what DoD requires (as evidenced by the blank spaces in the table). For example, the Navy designated its OPSEC, CoC, and substance abuse courses as "recommended," thus allowing unit commanders to determine who receives these courses and how often participants receive them. The Coast Guard falls under DHS during peacetime and is not subject to DoD guidance for GMT. It is not clear why some services do not have training for all topics even though, according to our assessment, each GMT topic is required. The Navy has recently made significant reductions in its GMT requirements and combined some topics. Further reduction is not a likely alternative; in fact, an increase may be called for.

We examined the GMT content contained in CBT on the services' LMSs. An overall observation is that, although training is presented differently, the overall training content appeared to cover the same material.

Overall, the estimated time to complete GMT per service requirements is the least in the Navy (three hours) followed by the Coast Guard (3.25 hours) and Air Force (6.25 hours). The Navy does not require as many GMT topics as the other services, and the number of Coast Guard topics is lower as well. The Army (nine to 16 hours) and Marine Corps (13.5 hours) devote significantly more time to GMT requirements. Most of the training is available to be done by CBT; however, as noted earlier, preferred delivery methods vary.

Summary

Overall, our assessment of service GMT programs is that they have been developed independently of one another and therefore are largely unstandardized. The stove-piped approach to GMT means that there are substantial opportunities for standardization across services. Within the Army, for which much of the training is locally prepared, there is potentially a great opportunity for standardization of training, which could result in improved use of time and resources.

Most of the services have made recent strides in examining and reducing the GMT burden. Most prominently, the Air Force and Navy have reduced GMT burdens. The Air Force has combined topics and directs that most of the training be done online by means of its LMS. The Navy has recently made significant reductions in its GMT requirements and combined some topics. Further reduction is not a likely alternative; in fact, an increase may be called for. The Marine Corps is closely examining its GMT requirements with the goal of streamlining training demands.

CHAPTER FOUR

Why General Military Training Is a Challenge

We examined DoD and higher-level directives, such as U.S. Code and presidential directives, to determine the source of training requirements. In Chapter Two, we listed the guidance that directs GMT. Mandated training requirements are often indicated annually or to be conducted "periodically throughout a service person's career." These mandated GMT requirements create demands for individuals' (and groups') time that directly compete with time needed for an individual's specialty training, mission (group and collective) readiness training, and maintenance and readiness demands.

Although GMT provides important knowledge, there are challenges and opportunity costs associated with completing it. Our sponsors asked us to examine GMT requirements because completing them takes participants' time and because these requirements can conflict with other RC premobilization unit and collective training demands. Service SMEs said that finding time to do GMT is a challenge for both the RC and the AC. In this chapter, we address the scope and nature of GMT challenges—the number of personnel who need the training, the amount of time required to do the training, AC compared with RC time demands, and GMT guidance and oversight needed.

The Number of Personnel to Train Is Significant

All uniformed service personnel—both AC and RC personnel—must complete GMT requirements prescribed by their services, so the audience is large. There are opportunity costs associated with the high GMT demands. When uniformed service members are receiving GMT, they are not available to perform their missions, mission training, specialty training, or collective training. A reduction in cyclic mandatory GMT requirements can reduce premobilization training burdens on the RC and allow personnel to maximize time available to meet mission requirements. To clarify the scope of training demands, Figure 4.1 illustrates the number of uniformed personnel, by service and component, who are required to complete GMT. The aggregate number of uniformed personnel is approximately 2.2 million. Because the number of personnel required to receive the training is high, small or marginal changes to GMT time demands will have a large effect as the changes are multiplied across the force. For this reason, governance of GMT topics and time requirements is important. An increase (or decrease) in GMT affects the entire force.

Figure 4.1
Number of U.S. Uniformed Personnel, by Service and Component, Fiscal Year 2009

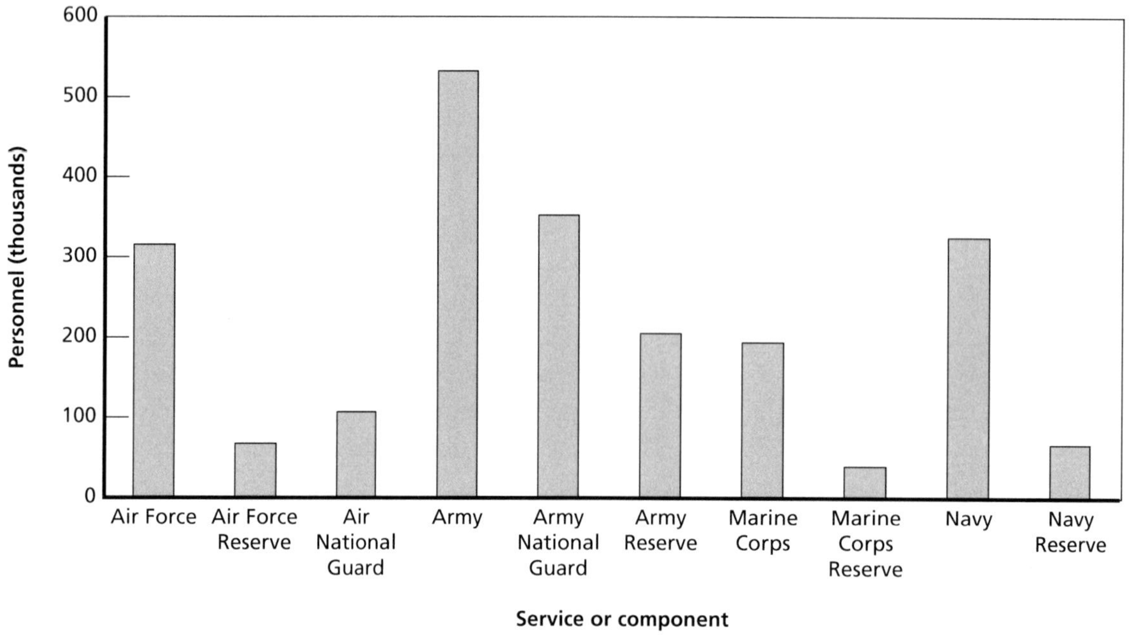

RAND *TR1222-4.1*

The Amount of Time Devoted to General Military Training Is High

In Chapter Three, we presented the amount of time required to complete GMT per individual. The notional time required for completing GMT varied by service. For the Army, the time demands ranged from nine to 16 hours; Navy, three hours; Air Force, 6.25; and the Marine Corps, 13.5 hours. We multiplied the number of notional hours needed to complete GMT by the number of personnel in each service or component to get a sense of the overall time demands that GMT places on uniformed personnel. Figure 4.2 illustrates these calculations. For the Army, the time shown in this figure represents the low range of time (nine hours per individual). The aggregate time required completing GMT across the DoD amounts to millions of man-hours.

In the aggregate and as represented in the figure, the services collectively devote more than 17.2 million man-hours annually to GMT. The Army and Marine Corps have the largest numbers of hours required to do the training. The Army has the most personnel, and, when combined with requiring the most time to do the training, the amount of time the Army devotes to GMT in the aggregate is largest of any service by far. The Navy's recent reductions in GMT have reduced its time demand considerably for both the Navy and the Navy Reserve. The Air Force's focused approach to reducing ancillary training (currently 6.5 hours estimated per individual) pays dividends when multiplied across the force. The Marine Corps, although it has the fewest personnel in uniform, expends more than 2.5 million man-hours completing GMT. The Marine Corps Reserve is the smallest RC, but its GMT requires more than 500,000 man-hours to complete annually.

Figure 4.2
Total Number of Hours Needed to Complete General Military Training, by Service and Component

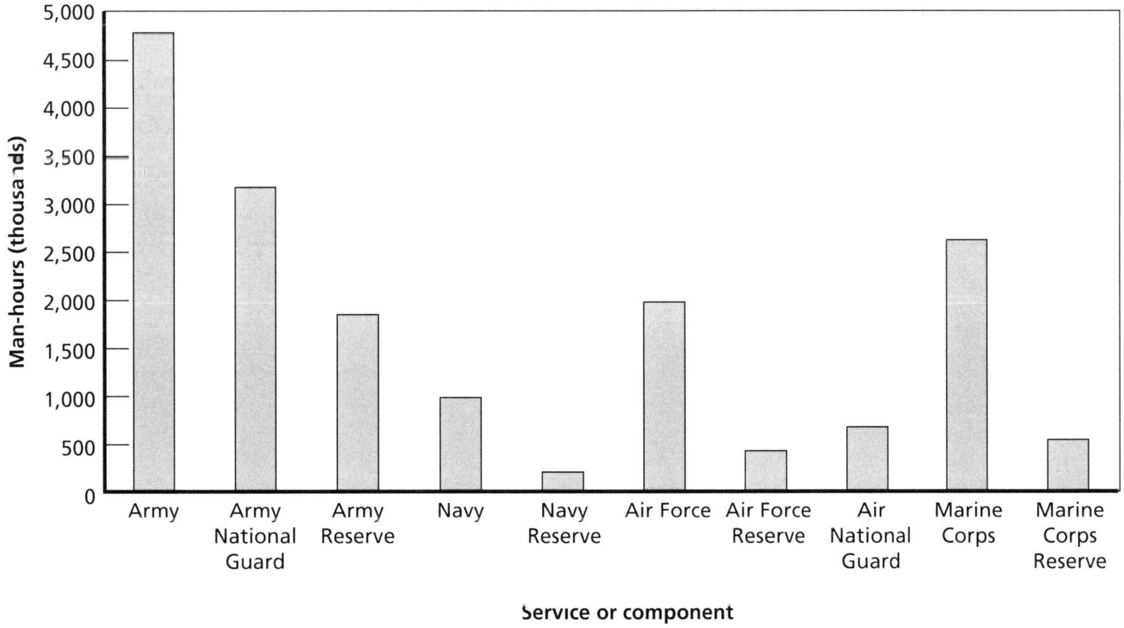

RAND *TR1222-4.2*

As we noted earlier, the AC and RC have competing demands for their time. Personnel must gain proficiency in their military occupational specialty (MOS), perform individual and collective training, and complete mission needs (e.g., maintain or prepare equipment for operation). The many demands on uniformed personnel's time must be balanced. As we learned through our discussion with SMEs, commanders place their primary emphasis on using their personnel's time to meet mission requirements. Service officials indicate that there is little time left over to meet GMT requirements, but they need to be met.

Reserve Component Personnel Have Less Time Available for General Military Training Than Active Component Personnel Have

Although both the AC and RC indicate that the available time is short, the percentage of time expended on mandatory GMT requirements consumes a substantial portion of an individual RC's available training time as compared with that of AC personnel.[1] As noted earlier, the amount of time required to complete GMT varies by service. We estimated the time to complete GMT versus the time available by uniformed personnel. Figure 4.3 illustrates the percentage of time that mandatory GMT consumes as a percentage of available training time for AC personnel. Although the percentage of time needed to complete GMT for AC personnel

[1] Service SMEs indicated that RC GMT requirements are normally completed during weekend drill periods (i.e., during the two eight-hour days per month). RC annual training periods (approximately two weeks in duration) are focused on specialty training and meeting mission demands (group and collective). We estimate weekend drill periods to amount to 192 hours of available time per year. For AC, we estimate 2,080 hours (52 weeks times 40 hours per week) available per year.

Figure 4.3
Percentage of Available Training Time Needed to Complete General Military Training, by Service, for the Active Component

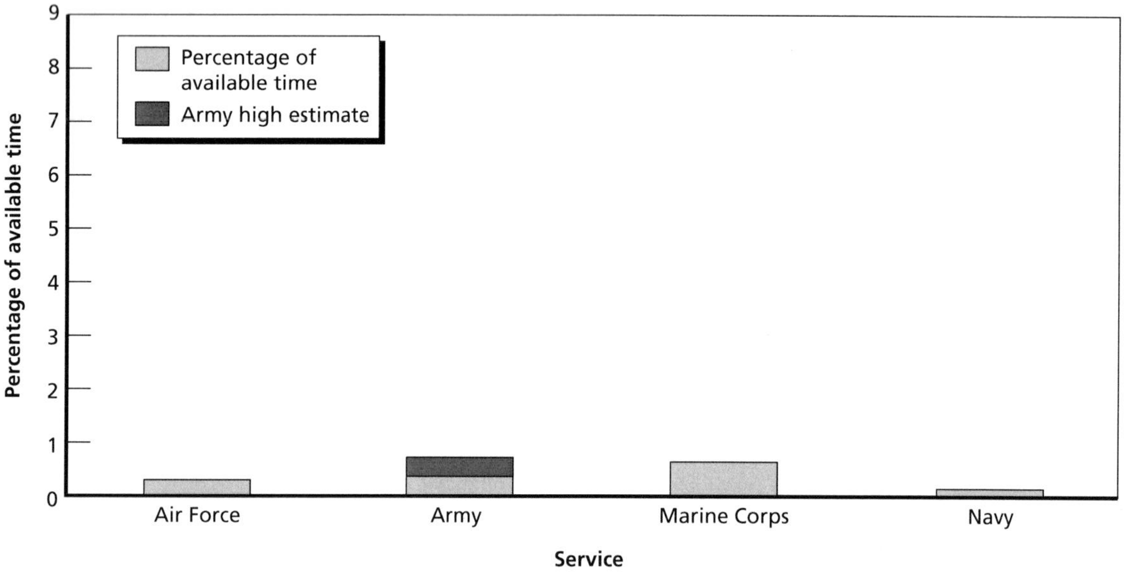

RAND *TR1222-4.3*

is small, AC SMEs indicated that it is still a challenge to accomplish and that trade-offs occur when GMT must be done.

The percentage of time needed for GMT as a fraction of total training time available needed by the AC is small. However, the percentage of time needed to complete GMT by all the RC is higher and, of course, greater than that of the AC because the RC has less time available in which to do it. Figure 4.4 illustrates the percentage of time that mandatory GMT consumes as a percentage of available training time for RC personnel.

Time demands for training vary by service. The Army gave a range of time needed to complete GMT (the Army notional estimate is nine to 16 hours annually). The Army National Guard (ARNG) and Army Reserve time demands needed to complete GMT requirements can vary, as depicted in Figure 4.3, from approximately 5 to 8 percent of available time. The percentage of time needed by the Marine Corps Reserve is 7 percent annually, followed by AFR and ANG (about 3 percent) and the Navy Reserve (about 1.5 percent).

Appropriate balances must be struck in meeting mission demands and GMT. Opportunity costs of completing GMT directly affect time available for mission training, particularly with the RC. Effective, efficient training that is standardized, universally accepted, and shorter can return time back to the unit commander to meet mission needs.

General Military Training Is Not Standardized in Some Cases Within a Service or Component

Army, ARNG, Army Reserve, and Marine Corps SMEs, noted that GMT is often developed locally. That is, instructors designated by the unit to provide the GMT to unit personnel develop the training on their own.

Figure 4.4
Percentage of Available Training Time Needed to Complete General Military Training, by Service, for the Reserve Component

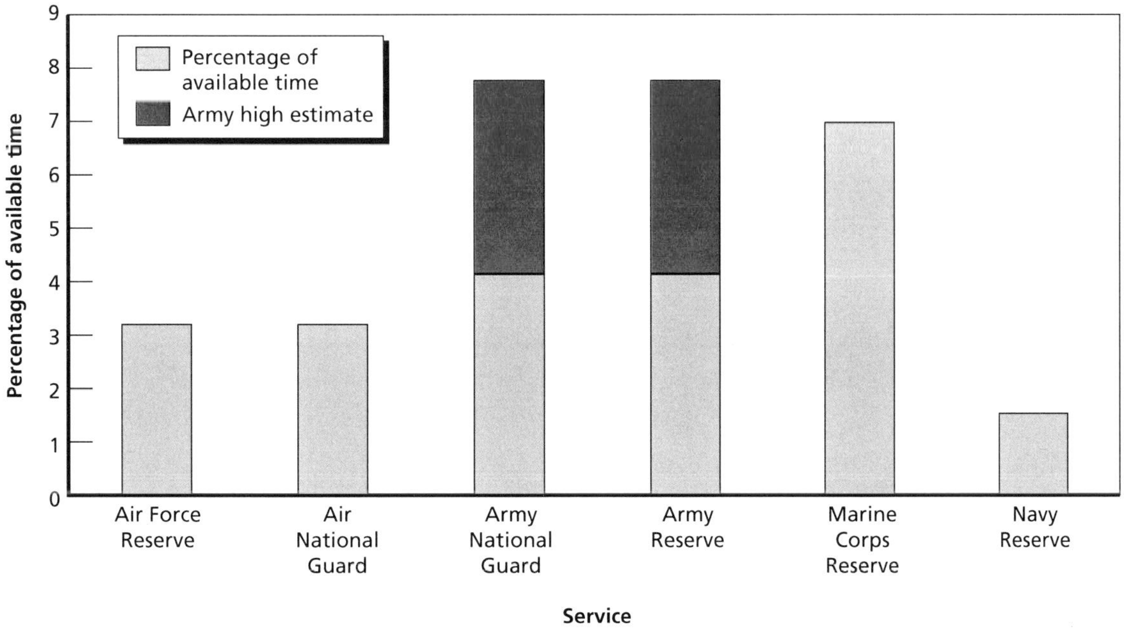

RAND *TR1222-4.4*

The Air Force and Coast Guard use of CBT as the primary delivery method supports standardization of GMT content throughout the Air Force and Coast Guard. The Navy prefers and utilizes a command presentation format (or instructor-led training). Standardized instructional material (instructor guides) is downloaded from the Navy's LMS by all Navy instructors who provide the training, so the Navy uses standardized instructional materials.

The Army and Marine Corps prefer the stand-up delivery method. Therefore, many instructors across the force, for both the AC and RC, develop local training materials for presentation to troops. Both the Army and Marine Corps could realize efficiencies, in terms of instructor preparation time and standardization of content, by maintaining a site from which to download instructional materials and instructor guides. These materials could be made available on their respective LMSs and would promote standardization within the service.

General Military Training Development and Sustainment Consume Resources

In addition to devoting valuable training time to GMT topics, each service independently expends financial resources to develop, prepare, and deliver its own GMT program. The potential exists for cost avoidance by preparing common, standardized curricula and CBT modules that could be used across the services.

Computer-Based Training Can Be Costly

Table 4.1 contains some service estimates of the cost of preparing computer-based GMT. These costs represent one-time development costs for CBT modules. Most services have transitioned

Table 4.1
Service Estimates of the Cost of Preparing Computer-Based General Military Training

Product	Cost ($)
Ten one-hour training lessons	500,000
Three 30-minute training lessons	50,000–150,000
One one-hour training lesson (average)	23,000
One 15-minute suicide prevention video[a]	890,000

[a] Not representative of the service's training costs.

from developing training to sustaining and maintaining training. Service SMEs whom we interviewed reported that postdevelopment sustainment costs are far lower than development costs. Overall, the estimated average cost of preparing GMT is approximately $23,000 to $50,000 per training lesson. The last entry in this table (the $890,000 estimate) was the amount to fill a short-fused need for a suicide prevention video. Although this was not considered a representative cost, resources were expended to purchase this short-fused training requirement.

The services do not routinely use other services' GMT CBT. An optimal method of conserving resources is to develop standardized CBT, paid for once, that can be used by all the services.

We did find a few instances in which one service chose to use another service's CBT rather than independently finance and develop its own training package. Also, one GMT topic, IA, produced by the Defense Information Systems Agency (DISA), is required training for all the services. The IA training is completed online as CBT. Table 4.2 is a summary of CBT shared by more than one service.

Stand-Up Training Is Also Costly

Up to this point, we have focused on the financial costs of CBT because those are easier to quantify than those for classroom (i.e., stand-up) training. During our interviews, the services reported that there is a significant amount of time spent preparing for and attending stand-up instruction sessions. Stand-up training costs include the resources expended preparing the GMT material (either by an individual or at a central location), trainer costs (time) for familiarizing him- or herself with the material, assembling the participants, delivering the training, checking comprehension, and recording the training.

Depending on the service, either training curriculum is developed locally at the unit level or a central program manager at the service headquarters level prepares instructor guides. Additionally, some GMT courses (e.g., EO, sexual harassment prevention, and suicide pre-

Table 4.2
Computer-Based Training Used by More Than One Service

Training	Producing Organization	Consuming Service
IA	DISA	All DoD, Coast Guard
Law of land warfare	Army	Marine Corps
OPSEC	Army	Marine Corps

vention) require trainers who themselves have attended specialized training to be certified to be a trainer, resulting in additional costs. According to discussions with SMEs, the recurring travel and personnel costs for this process are likely to be much higher over time than for the CBT courses. However, we found no evidence that the cost-effectiveness of either stand-up or computer-based GMT has been evaluated.

Stand-up training preparation, delivery, and checking comprehension take time, and recording training accomplishment also adds administrative time. When considering that GMT is given across the entire force, the costs of stand-up training can be significant and likely are greater than those of CBT. With CBT, the training is available anytime and anywhere (to those with access to a computer), and training completion is recorded electronically. However, as we noted, the services have preferred and prescribed delivery methods; some prefer stand-up training, and all want a variety of delivery options.

Secondary Effects of General Military Training Also Burden Units

Completion of GMT must be tracked and recorded. Some GMT must be reported to higher authority, which requires individual training to be tracked, recorded, and reported by the unit as completed. Tracking and reporting creates additional time and administrative burdens, unless done online by means of a service LMS. Not every service wants to conduct training online by means of LMS, and it is infeasible for delivery of training to all uniformed personnel.[2]

Summary

GMT demands are many; the aggregate effect on uniformed service members in terms of time demands across the force is high, especially on the RC; and guidance that directs that training be completed comes from various DoD directives. Although the AC has more time than the RC has available to complete GMT, SMEs indicated that the opportunity costs of completing GMT also must be balanced with mission demands. The training and time demands of this magnitude need careful management because small changes can yield big rewards or impose large burdens.

GMT is not standardized across the services or, in some cases, within a service. There are also secondary effects of GMT that burden units, including recording attendance and reporting GMT completion to higher authority.

---

[2] The Coast Guard, however, successfully uses an LMS to train and track GMT accomplishment.

CHAPTER FIVE

What Options Exist to Standardize Requirements for General Military Training and Reduce Its Burden?

This research identifies the baseline of mandatory GMT topics common to all the services. OSD also asked RAND to identify strategies to standardize requirements and reduce the GMT burden with the goal of optimizing available training time and training effectiveness. In this chapter, we examine and discuss potential opportunities to standardize training requirements and reduce the burden of GMT. Note that we have not done an independent analysis of the options discussed. Should OSD wish to adopt any of them, it would need to analyze them to determine whether they produce the desired outcome and do not have unintended consequences. We offer them as examples of what has been tried to reduce the burden of GMT and make it more efficient.

Options at the Service Level

We compared service GMT programs and relevant literature to identify possibilities to standardize GMT requirements and reduce GMT burden. The motivation for standardization is not to force the services to conform GMT to a single approach; rather, it is to address training objectives and desired outcomes and to reduce redundancy. The benefits of standardized training are that common instructional materials can be prepared and used across the board. Standardized instruction would eliminate the need for the services to develop and update training independently. Standardized instruction could be developed once, and the core training materials shared among the services. These core instructional materials could be tailored by the services, as needed, to current scenarios, situations, or locations. Standardized instruction promotes universal acceptance of completed training, at individual and joint commands.

Chapter Four detailed the time demands that GMT places on AC and RC personnel. The benefit of a reduction in GMT time requirements is that it would increase an individual's time available to engage in mission- or specialty-oriented training or perform other assignments. A major goal in the reduction of GMT time requirements is to give time back to the unit commander to use unit personnel as he or she sees fit.

Services' Current Efforts

The review of guidance directing GMT indicates that the services' GMT programs derive from common guidance. The benefits of standardized content are that all personnel are trained to a common standard, training done at one unit is transferable to another, and there is no need to prepare training materials locally. Standardized content could be used as the baseline training, and each service could tailor and supplement the training based on the needs of the trainees and as scenarios and situations dictate. The services could provide differentiation training as required.

Multiple services (including the Coast Guard) have made an effort to standardize or reduce each of the 14 topics. IA is the best example of standardized training across the board: The training content is standardized, and all uniformed personnel must take the same training. However, this approach is the exception rather than the norm.

Attempts have been made to conserve resources and promote training standardization through the establishment of training "repositories." However, research indicates that potential users have strong incentives not to use such repositories and that those who do use them encounter a host of technical problems (e.g., the training does not run well on other organizations' LMSs) (Shanley, Lewis, and Straus, 2009). These obstacles will need to be addressed.

The services have made other efforts to standardize requirements and reduce the burden of GMT, and Table 5.1 illustrates examples of these efforts. We shaded in training topics, by service, for which standardization or reduction efforts have occurred. In addition, we include textual indications of the services' actions. For example, the Navy changed some GMT from annual to highly encouraged training, which places discretion over these topics with the commanding officer.

IA training is provided by a central DoD source and cannot be reconfigured or recombined by the individual services. This curriculum was developed with the support of DADLAT, and the standard DISA IA training is done online. DADLAT personnel with whom we spoke indicated that having them develop this training was a rare opportunity and that they are not resourced or staffed to produce standardized training. The Army, Navy, and Marine Corps use the standardized CTIP training in an online training format.

The Navy combined EO, sexual harassment, and SAPR training into one training segment. Unit personnel provide this training by means of stand-up instruction. Navy commanding officers are "highly encouraged" to establish recurring training programs for OPSEC, CoC, and substance abuse (drug abuse prevention and control and tobacco use prevention and cessation).

The Air Force reduced airman training demands by shifting some training to be event-driven (CI, substance abuse, and EO), combining related training (OPSEC, NATO security, and Privacy Act), and reducing some training audiences to only a selected force. The Air Force also combined related curricula in TFAT. It evaluated each topic to determine whether it needed to be conducted annually and then extended the periodicity for many topics by giving airmen training at different points in their careers, such as at accession sites or the schoolhouse. The Air Force also reduced training time by evaluating topic time durations and scaled back some time requirements. It replaced nine blocks (or topics) of instruction with three blocks. The new blocks are information protection, FP, and human relations. The blocks that they replaced are

- information security
- NATO security
- IA
- record management
- Privacy Act
- Freedom of Information Act (FOIA) (5 U.S.C. § 552)
- protection from terrorism

- CTIP
- suicide awareness.

By combining related training, the Air Force estimates that TFAT will save 6 million man-hours per year (Thompson, 2007).

Table 5.1
Examples of Service-Initiated Standardization and Reduction of General Military Training

Topic	Army	Navy	Air Force	Marine Corps	Coast Guard
AT					Combined with CI awareness and OPSEC
CI awareness			Event-driven	Use Army training	Combined with AT and OPSEC
CoC		Highly encouraged		Use Army training	
CTIP	Standard	Standard		Standard	Standard
EO		Combined with sexual harassment and SAPR	Event-driven		
IA	Standard	Standard	Standard	Standard	Standard
NATO security			Combined with OPSEC and Privacy Act/PII		
OPSEC		Highly encouraged	Combined with NATO security and Privacy Act/PII	Use Army training	Combined with AT and CI awareness
Physical fitness	Service specific	Service specific	Service specific	Service specific	Service specific
Privacy Act/PII			Combined with OPSEC and NATO security		
SAPR		Combined with EO and sexual harassment			Combined with sexual harassment
Sexual harassment		Combined with EO and SAPR			Combined with SAPR
Substance abuse prevention		Highly encouraged	Event-driven	Combined with suicide prevention	
Suicide prevention				Combined with substance abuse prevention	

NOTE: Green indicates a training topic that has been standardized. Maroon indicates a training topic that has been combined with or is completed together with another topic. Blue indicates a training topic that is now event-driven (e.g., required upon reenlistment). Gold indicates training for which one service uses another service's training plan. Red indicates a training topic that is service specific and thus is not considered for standardization or reduction. White indicates a topic for which we found few or no standardization or reduction efforts. The Coast Guard falls under DHS during peacetime and has different GMT requirements. We include the Coast Guard to illustrate efforts it has taken to standardize or reduce training.

The Marine Corps has adopted the Army training for some topics (CI, OPSEC, CoC) and combined some topics (suicide prevention and substance abuse). The Coast Guard combines security-related training topics (AT, CI, OPSEC) and sexual harassment and SAPR. Each service has service-specific physical fitness training.

Overall, the Army has made little effort to reduce GMT, even though it potentially has the most to gain from reduction efforts. The Navy has made drastic cutbacks in training that reduced GMT from 11 to five topics annually (and may have gone too far because some topics that are, according to our assessment, required are not included in the Navy's GMT plan). The Air Force has taken a measured approach by reducing the training needed by all, tailoring training to only those who need it and by making some "periodic" training requirements event-driven.

Service-Level Options

We discussed these options at an OSD-sponsored meeting with service SMEs. Potential options to standardize GMT requirements and reduce the GMT burden include the following:

- Combine the curricula of similar training topics.
- Extend training periodicity.
- Narrow the audience scope by mandate or exam.
- Reduce the training content.
- Produce common CBT and instructional material.
- Periodically review training content.

In their efforts to reduce the GMT burden, some services have already implemented some of these options. However, their efforts have been largely independent of one another and have not led to standardized instruction among the services. The standardization and reduction of the burden of training required of all uniformed service personnel can lead to greater efficiencies. We discuss and describe these options and the services efforts in this section.

Combine the Curricula of Similar Training Topics. Combining similar individually mandated training topics into one training curriculum is a popular option. It enables services to reduce overall training times by eliminating some of the duplicative introductory or explanatory materials, and it reduces administrative or overhead time in assembling personnel for the training. Several approaches have been taken to combine training topics. Some of the low-hanging fruit, i.e., topics that are easy to combine, are those that address the respectful treatment of others. Along these lines, both the Navy and the Coast Guard have combined sexual harassment and SAPR training into one topic. The Navy also included EO in this same training curriculum.

There appear to be competing philosophies for combining the security or information training topics. The Air Force consolidated several "how to handle information" training topics, such as OPSEC, NATO security, and the Privacy Act, while the Coast Guard took a different approach and combined several subjects with a security awareness perspective into one package (AT awareness, OPSEC, and CI awareness).

The Marine Corps combined substance abuse and suicide awareness training into its Semper Fit program. This program is a holistic, "whole-marine" wellness program designed around maintaining a healthy mind and body. Regardless of the approach implemented, each of these approaches has the same effect: reducing the total time spent training on these topics.

Setup and administrative time is needed to deliver GMT, e.g., taking roll and signing muster sheets. By combining topics, rather than doing them separately, setup and administrative time is reduced. The more topics that can be combined, the less administrative time is needed per GMT topic.

Extend Training Periodicity. Adjusting the periodicity of training requirements is another approach that could be implemented to reduce the training burden. Nearly all the GMT topics conducted by the services are completed on an annual basis (with some exceptions). However, DoD could extend the periodicity of GMT demands or use a triggering mechanism other than an annual calendar. DoD guidance states that training is "required upon initial entry, periodically, and throughout career" for EO and human relations, sexual harassment, and substance abuse training. The Air Force has already changed the periodicity of substance abuse, CI awareness, and sexual harassment training. These topics are now triggered by certain events, such as promotion, reenlistment, assignment to a service school, or transferring to a new unit. Similarly, the Navy converted its homosexual conduct policy training to event-driven training in an effort to reduce periodicity.[1]

Narrow the Audience Scope by Mandate or Exam. Narrowing the scope of training is another approach that can reduce the overall training burden. The Air Force narrowed the scope of the homosexual conduct policy training to a selected subset of individuals (supervisors). This training teaches them how to handle the issue when it arises. The Air Force also has selected force ancillary training, which includes the Notification and Federal Employee Antidiscrimination and Retaliation Act (No FEAR Act), survivor assistance awareness, homosexuality policy, ethics, and the Lautenberg Amendment.[2]

The Navy designated its OPSEC, CoC, and substance abuse courses as "recommended," thus allowing unit commanders to determine who receives these courses and how often they receive them. This type of scope-narrowing option could also be accomplished for senior personnel who are already expected to be well versed in many of these topics, having attended the same courses annually throughout their careers. If the audience scope is narrowed by mandate, the service is assuming that the individuals who fall outside the scope either already know or do not need the training content. If there is uncertainty about individuals' level of knowledge of these topics, especially for senior individuals, a pretest may be used to evaluate the level of knowledge. This step reduces retraining of knowledgeable individuals and allows training time and resources to be focused on individuals most in need of training.

We noted in an earlier section that the Coast Guard currently uses a pretest for its CBT modules. The Coast Guard has a goal of implementing this type of pretest for all its training. Once implemented, all computer-based GMT will be customized by an individual's demonstrated level of knowledge within a GMT topic on a section-by-section basis.

Reduce the Training Content. Another option we have identified to reduce the training burden is to reduce the total amount of material included in the training curriculum. We have not observed any service-specific examples for this option. However, we have observed that DoD has not specified any "minimum" curriculum for any of these training topics. One

[1] A recent revision to law repealed 10 U.S.C. § 654, policy concerning homosexuality in the armed forces.

[2] The No FEAR Act enforces antidiscrimination and whistleblower protection laws. The Lautenberg Amendment makes it a felony for anyone convicted of a misdemeanor crime of domestic violence to ship, transport, possess, or receive firearms or ammunition. It is also a felony for any person to sell or otherwise dispose of a firearm to anyone known to have such a conviction. ("The Lautenberg Amendment," undated)

way to facilitate service standardization and potential reductions in GMT total training time would be for DoD to promulgate core training objectives or curriculum content for all GMT topics. This would allow services to compare their current training packages with the minimum requirement and make adjustments as needed.

Produce Common Computer-Based Training and Instructional Materials. The use of high-quality CBT provides an option that can be used by both AC and RC. SMEs want options, and pursuit of this strategy adds an option for more standardization and reduction of the GMT burden. High-quality CBT can reduce training time, as noted earlier. The Air Force and Coast Guard use CBT to a great extent to meet GMT demands. All of the Coast Guard GMT topics are available by means of CBT, as are many Air Force topics. The Army, Navy, and Marine Corps focus their delivery on stand-up instruction and likely do not maintain or update the CBT GMT topics to the extent that the Air Force and Coast Guard do. Going forward, a standardized common GMT CBT option that could be used across DoD could and should be pursued by DoD for the use of all the services.

The development and sharing of common CBT and instructional materials is an option for both standardization and reduction of the training burden. As noted in Chapter Three, CBT has many advantages over stand-up training delivery, including standardization. Although there is no one best way to deliver training, the availability of options for training delivery has benefits. The benefit of making options available is that it accommodates the fact that people learn in different ways and at different speeds. Although some GMT (e.g., DoD-standard IA training) can be delivered in only one way, other topics can be delivered in multiple ways. There also may be some accommodations needed to meet the various constraints of target audiences, for which stand-up training may present challenges. For example, it may be challenging or time-intensive for instructors to provide instruction for stragglers who miss the training session.

Often, the GMT instructional training material is developed locally. The Army and Marine Corps could realize efficiencies, in terms of instructor preparation time and standardized content, by maintaining a site from which to download GMT instructional materials and instructor guides. These materials could be made available on their respective LMSs and would promote standardization within the service and across the services. The Navy currently has a centralized GMT website from which to download instructional materials. The Army and Marine Corps could use a similar approach.

CBT reduces the training burden on the individual by giving him or her (and the unit) more choices to complete required training. CBT versions of most GMT exist now. Although CBT may not be some services' preferred way to deliver training, it is an option that has benefits. Providing an option of taking the training by means of CBT may be an effective and efficient solution for those who missed the training. In addition, completing GMT via CBT may be an acceptable alternative for the RC, in which individual and unit training and mission demands need to be carefully balanced with GMT demands.

Additionally, there are numerous other topics that some members of the uniformed population must receive that fall outside the definition of GMT. Nevertheless, this training is required for a significant portion of the service population, and CBT may be a cost-effective and efficient method of delivering this training. For example, the articles of the Uniform Code of Military Justice (UCMJ) must be carefully explained to each enlisted service member upon his or her reenlistment.[3] There are many articles, and it may be cost prohibitive for a qualified instructor to provide this training to each reenlister. The use of standardized CBT may be a very appropriate approach to conducting this training.

Periodically Review Training Content. An option we have identified to reduce the training burden or to make the GMT relevant is to periodically review training content. We did observe service-specific examples of this option. Updating or removal of irrelevant or outdated GMT content is essential to an effective GMT program and to keep the audience engaged. Reviewing training content would include a comparison of current GMT packages with the current situation or environment and then tailoring the training by making adjustment as needed.

Effective programs include locally or culturally relevant content. Service representatives, the Army in particular, discussed the benefits of instructor-led training because it allows the tailoring of training to be more specific to the current time or location. For example, CTIP GMT could be geared to increase service members' awareness to the current and specific locations and CTIP practices in the area in which they are operating.

The review of training content makes the training relevant, puts the training in an appropriate context, and makes it meaningful to the audience.

U.S. Department of Defense–Level Options

The examples provided in the previous section indicate approaches that one service has taken that could be applied to other services. In addition, DoD and the services could take action to standardize training across the board.

Produce or Adopt Common Computer-Based Training

SMEs and a DoD OPR in charge of training required for all uniformed members should evaluate the quality of the content and instructional design of current GMT by CBT. Production and adoption of CBT should be investigated to find opportunities to provide standardized training that can be taken in less time. Collaborating on new and improved CBT could be an important way to engage stakeholders so that the training meets objectives and is generalizable across the force.

Use Existing Resources That Can Address Training Challenges

DoD conducts a great deal of interservice training. Interservice training is conducted by consolidating or colocating training courses that are similar in content across the services. A lead organization that guides interservice training is the Interservice Training Review Organization (ITRO). ITRO is made up of representatives of all the uniformed services. ITRO coordinates the multiservice training for common training requirements, which enables the services to preserve training resources.

The ITRO training study process is intended to facilitate the establishment of consolidated and colocated training solutions. The process provides a foundation for the analysis of

---

[3] Training on the UCMJ upon reenlistment applies only to the enlisted force per 10 U.S.C. § 137.

training requirements, the development and implementation of training, and the review of existing curriculum.

The ITRO primarily focuses on technical training courses that are performed across the services—for example, enlisted medical training. Much of the training that enlisted medical specialists need is similar, regardless of their parent service. The ITRO uses procedures to evaluate and analyze the cost-effectiveness of consolidation (and colocation) of training. ITRO goals could be considered and used by DoD in efforts to standardize and reduce GMT:

- Consolidated training should create savings or increased training effectiveness or efficiency through the development of common core curricula and use of common equipment or facilities.
- Consolidation or colocation approvals should be evaluated on the basis of DoD cost and savings, improved training effectiveness and efficiency, and overall resource savings. A long-term, cost-effective action is the objective; however, improvement in training efficiency may offset increased costs.

The ITRO and its training study process may serve as a vehicle to support increased standardization and reduction of the GMT burden.

The ITRO's mission is to achieve training efficiencies through consolidation or colocation of common training among the services (Society of American Military Engineer, 2006). Although the ITRO's mandate is to focus on courses and skills that are common among the services, its methodology could be easily adapted to producing a common GMT curriculum or set of CBT courses.

DoD has extensive experience with consolidating training with an eye toward standardizing its content and thereby reducing the level of resource input required to produce training materials or conduct training. The ITRO's standardization and reduction processes have worked historically well for entry-level training but not so well for higher-level tasks. GMT fits nicely into the former category.

However, fostering service cooperation in producing standardized training is not without its pitfalls. If a service's parochial interests and cultures are not sufficiently represented, that service is not likely to incorporate jointly produced training materials into its training program (Viccellio, 1995). Indeed, gaining the services' agreement on a common training curriculum can be challenging because the roles, missions, and scenarios under which each service operates differ. Service training SMEs indicated that standardized GMT must be relevant to each of the services' personnel and that standardized training must be designed in a way that all recipients from all services can relate to it.

Provide a Gatekeeper

If new training requirements are levied across DoD, the magnitude of GMT requirements can grow, forcing unit commanders to make additional decisions between mission training demands and GMT-mandated requirements. The assessment given earlier indicates the number of personnel and man-hours and the percentage of time required to complete GMT. New training requirements are additive to existing requirements, have a big effect (i.e., increase the time required to complete GMT), and are multiplicative across the force. For example, eliminating or adding a half-hour annual GMT topic could reduce or add to the time needed for training across the DoD force by 1.1 million man-hours annually and incur training development costs. The opportunity costs of GMT suggest a need for governance or a gatekeeper role that assesses or limits the effects of new GMT requirements to be levied on the force. This role should be established at the DoD level. The goal would be to assess training requirements, increase effectiveness and cost-efficiency, resolve issues, balance priorities, and gain training efficiencies.

The Air Force has adopted this approach and has a detailed screening "gatekeeper" process that is used to vet new GMT requirements. The Air Force recognized the toll that GMT takes on its airmen's time and established a gatekeeper process for vetting GMT requirements. The purpose of this process is to ensure that senior leadership has an awareness of the time it takes to complete training, the toll it takes on the force, establishes training priorities, set limits, and communicates results. These positive steps in assigning responsibilities by means of a gatekeeper role are a method to monitor training and limit training demands. Although a gatekeeper role is performed at the service level, broader oversight across DoD may be warranted.

High-Level General Military Training Oversight Is Absent

The training requirements for GMT are contained in multiple DoD directives and instructions that mandate GMT for the forces. There is little oversight and control of GMT. A centralized DoD Office of Primary Responsibility (OPR) could be tasked with determining, evaluating, validating, and coordinating GMT requirements that must be completed by all uniformed personnel across the force. Currently, there is no OPR that performs these responsibilities. There are offices responsible for readiness and training, under the Office of the Under Secretary of Defense for Personnel and Readiness (OUSD[P&R]), that could potentially assume these responsibilities.[4]

Some services have made significant reductions in GMT, and some of the services may have gone too far. Our review of service guidance indicates that some topics that should be included in service GMT plans are not included. For example, the Navy has reduced annual GMT to five topics, and some topics that are required to be done annually are not listed in the service's guidance. As such, GMT requirements appear not to be implemented as intended.

If DoD is concerned and would like to standardize and reduce GMT, it must first understand the challenges, which this research serves to show. GMT places training and time demands on all uniformed members. To take steps to standardize and reduce GMT, the training needs to be under control. Control can take the form of delineating the requirements and publishing them in a single DoD directive or instruction. This guidance then would be a single source document that provides the GMT topics, periodicity, and overall guidance and direction.

An overall DoD directive could stipulate the content, tracking, and reporting demands. Currently, GMT guidance is issued from disparate DoD offices. In addition, a single DoD OPR authority could establish the gatekeeper responsibility described earlier to evaluate and validate new training demands. Moreover, this OPR could challenge current training requirements—training content, validity, periodicity, and reporting—to ensure consistency across the board.

Currently, disparate OPRs are in charge of training requirements. DoD OPRs, responsible for the training requirement, need to be asked these questions:

---

[4] Per DoD's military training directive (DoDD 1322.18), OUSD(P&R) is responsible for overseeing and providing policy for individual military training programs for the total force.

- Is this still a valid training requirement; if so, why?
- Does this training requirement need to be met on an annual basis? If so, why?
- Can the training periodicity be extended (e.g., done at key points in one's career) or given to selected audiences (only junior personnel, or mid-level, or supervisory)?
- Does the training requirement need to be tracked and reported? If so, to whom and why?
- What is the required training content? Should the services submit training content to the OPR for approval? Is there model curriculum that could be potentially adopted across the board?

A single DoD OPR needs to be in charge of GMT. This OPR, with an implementing directive, can validate or challenge training requirements, examine and determine methods to deliver effective and efficient training, standardize requirements, reduce the training burden, and serve as gatekeeper to validate and assess effects and make decisions for future requirements. In the absence of an OPR backed by an implementing directive, only ad hoc solutions are available to standardize and reduce GMT requirements.

What More Can Be Done to Standardize and Reduce General Military Training?

DoD is interested in starting down the path toward increased standardization and potential reduction of GMT, and it desires options that could potentially achieve this. We considered what the services have already done to standardize requirements and reduce the GMT burden as supporting criteria for suggesting wider potential changes.

Table 5.2 lists GMT topics, the periodicity of the topic (annual or periodic) from DoD guidance, and an assessment of the services' approaches to standardize the topic (made the topic event-driven, curricula used by more than one service, topic was combined with other topics, or the topic is service-specific). We list the categories of actions already taken by the services as criteria to consider further standardization or reduction options (across the top), and we annotate actions taken by the services compared with the criteria. In this table, purple indicates a topic that is standardized now, dark green indicates one that could potentially be standardized and event-driven, light green indicates a topic that could potentially be standardized, yellow indicates a topic that could be standardized but contains sensitive material, and red indicates a topic that should not be standardized. The research team examined these topics and what the services have done to standardize and reduce GMT training demands, then applied judgment in suggesting options to consider. For example, some training topics that are completed on an annual basis now by the services, e.g., EO, could be done periodically (following DoD guidance). The services have the authority now to conduct this training periodically or "event-driven" (e.g., when a service member reenlists), and they should do so as they deem appropriate. Extending the training periodicity reduces training demands.

Use of the same curricula by two or more services is the first step toward more-complete standardization. As noted, the Marine Corps uses Army training for some curricula. This suggests that standardization of these topics could potentially be extended to the Navy and the Air Force.

Table 5.2
How Potential Options Might Be Used to Further Standardize and Reduce General Military Training

	Criterion to Consider Standardization/Reduction Options					
GMT Topic	Periodicity Requirement	Event-Driven	Same Curriculum Used by Multiple Services	Topic Combined with Other Topics	Service Specific	Standardization or Reduction Option to Consider
AT	Annual			Coast Guard (CI, OPSEC)		Standardize
CI awareness	Annual	Air Force	Army, Marine Corps	Coast Guard (AT, CI)		Standardize
CoC	Periodic		Army, Marine Corps			Standardize, make event-driven
CTIP	Annual		Army, Navy, Marine Corps			Topic is standardized except for Air Force and Coast Guard
EO	Periodic	Air Force				Standardize, make event-driven
IA	Annual		Army, Navy, Air Force, Marine Corps, Coast Guard			Topic is standardized
NATO security	Annual			Air Force (OPSEC, Privacy Act)		Combine, standardize
OPSEC	Annual		Army, Marine Corps	Air Force (OPSEC, NATO security, Privacy Act), Coast Guard (AT, CI)		Combine, standardize
Physical fitness	Annual				X	Maintain as service specific
Privacy Act	Annual			Air Force (OPSEC, NATO security)		Combine, standardize
SAPR	Annual			Coast Guard (sexual harassment)		Engage OPR for standardization determination
Sexual harassment	Periodic	Air Force		Coast Guard (SAPR)		Standardize, make event-driven
Substance abuse prevention	None given	Air Force		Marine Corps (suicide prevention)		Engage OPR for standardization or reduction determination
Suicide prevention	None given			Marine Corps (substance abuse)		Engage OPR for standardization or reduction determination

Time savings and efficiencies can also be reaped by combining training topics, which the services have done for several topics, as noted in Table 5.1. Combining topics and providing instruction for two or more topics can save time.

Physical fitness training is mostly hands-on training, is done differently and tailored to the services, and should remain service-specific training.

We consider some training to be sensitive or needing additional expert input to determine the potential for standardization and reduction. These topics include suicide prevention, substance abuse, and SAPR.

AT awareness training is required for all service members, and it is a topic for which training could potentially be standardized. The Joint Staff has AT training available online (Antiterrorism Level I Training System, undated), and completion of this online training meets the annual requirement. This online training is standardized now; the website also offers downloadable PowerPoint presentations available for training officers to give lecture-based instruction. This training could be utilized now to standardize training across the services.

CI training is event-driven training for the Air Force;[5] the Army and Marine Corps use the same CI curricula, and the Coast Guard has combined this topic with OPSEC. The use of curricula by two services suggests the possibility of increased standardization across the board. A potential drawback of standardizing CI training is that individual servicemen and women may be targeted differently, necessitating service-specific and tailored training.

The Army and Marine Corps also use the same OPSEC training, and the Air Force has combined OPSEC with other security topics. Increased standardization of OPSEC training may be possible.

The Air Force has also combined NATO security and Privacy Act training with OPSEC. As noted earlier, the Air Force has achieved efficiencies in combining its security training (OPSEC, NATO security, Privacy Act). Potential may exist in combining training for these topics in other services. Linking like topics together can save setup time needed to get unit personnel or groups together to deliver required training and to record training accomplishment.

As noted previously, IA and CTIP are standardized to a large extent now.

CoC, EO, and sexual harassment training could potentially be standardized and made event-driven. Our research indicates that DoD policy dictates that these training demands be completed on a periodic, as opposed to annual, basis. Extending this training to be part of reenlistment training could significantly reduce the training demands for these topics and return time back to unit commanders.

These are steps that the services could potentially consider in future efforts to promote standardization or reduction of GMT demands. Table 5.2 is not an authoritative list but is one based on observed efforts across the services that could be extended to other services for consideration.

Standardized instruction could provide the minimum instruction that could be supplemented and tailored by the services to meet their needs. As noted earlier, some topics may not be appropriate for consideration of standardization or reduction efforts, and OPRs should be consulted before any standardization or reduction efforts are pursued.

Again, we note that we consider some topics (suicide prevention, substance abuse, SAPR) to be sensitive and needing expert input to determine standardization and reduction potential.

---

[5] DoD guidance directs CI training to be an annual requirement, but the Air Force CI training is event-driven. Therefore, the Air Force may not be implementing CI training as intended.

Summary

Options do exist to standardize and reduce GMT. Indeed, the services have recognized the challenges that GMT poses for their personnel and units and have taken action to reduce the training burden. They have done so by combining curricula of similar topics, extending training periodicity, narrowing the audience, reducing content, and using common CBT and instructional materials. There are many examples of actions that the services have taken individually to reduce GMT. A gatekeeper role may be needed to monitor training requirements being placed on the services because an additional single requirement has significant implications when applied across the total number of uniformed personnel. High-level oversight of GMT directed by DoD or above does not exist, and a DoD instruction that details GMT requirements and policy would reduce ambiguity and increase consistency among the services. DoD does have a history of interservice training and has an established organization, the ITRO, to find training solutions for common training. The goals and methods used by the ITRO are a model for efforts to standardize GMT requirements and reduce the GMT burden.

CHAPTER SIX

Conclusions and Recommendations

This chapter summarizes our conclusions and recommendations.

Conclusions

We drew the following conclusions from our research:

Time Devoted to General Military Training

Although GMT in the aggregate requires a substantial number of hours across the services, the amount of time as percentage of total training time for AC units is not high, constituting less than 1 percent of total training time. The burden falls more heavily on some RC units because they must accomplish the same GMT as the AC and have far fewer training hours available. The Army and Marine Corps RCs devote from 5 to 8 percent of available training time to GMT.

Standardization

Service GMT programs have been developed independently of one another and are not standardized. There appear to be opportunities for standardization, which could yield more-efficient use of time and resources.

Flexibility

All services want GMT that can be tailored to their specific purposes and that can be delivered in different modes to different audiences. An example would be CBT that could be used for makeup training for individuals who miss regularly scheduled training.

Reducing the General Military Training Burden

Several services have reduced the amount of GMT they require; however, in some cases, they have eliminated training that is still required. Some topics that should be included in service GMT plans are not. Additionally, our review of service GMT programs indicates that some GMT-required topics according to the approved working definition of *GMT* are not required across the board.

Need for Training

Much GMT has been a long-standing requirement. Training requirements should receive a careful review to determine whether they are still necessary and whether the current time requirements remain valid. Training time is finite, and every GMT demand means that some other training or mission demand is not possible. Thus, any new requirements should be scrutinized carefully.

Single Source of General Military Training Requirements

DoD GMT requirements are spread among many directives, which complicates the services' task to identify what DoD requires.

Potential Efficiencies

Both the Army and Marine Corps could realize efficiencies, in terms of instructor preparation time and standardized content, by maintaining a site to download instructional materials and instructor guides. These materials could be made available on their respective LMSs and would also promote standardization within the service.

Options

There are several ways to reduce the GMT burden and to make its delivery more efficient. Some services have already implemented options, and these appear to reduce time required for GMT and increase efficiency. However, these options have not been thoroughly analyzed.

Recommendations

In light of these conclusions, we recommend that DoD or the services take the following actions.

DoD and the services should consider using a DoD-wide accessible site that the services and components could use to download standardized GMT curriculum. To promote uniform acceptance and standardization (and potentially reduce costs), services' websites could link to the DoD-wide website. Service training management systems could then use the standardized training. A DoD-wide, accessible site could provide downloadable, flexible training options for unit commanders, such as lectures and online (CBT) training options. Compatibility issues between LMSs would need to be addressed for CBT instruction. Standardized GMT content would increase interoperability and training transferability.

DoD should consider adopting standardized CBT GMT. Collaborating on new and improved CBT could be a main way to engage stakeholders. The services have already taken steps to standardize and reduce GMT, and continued progress could be made through increased collaboration and sharing of ideas. Services could share what lessons they learned. Service SMEs desire options for training delivery. The sharing of information and best practices by all in the use of CBT can help the services provide well-designed, service-generic CBT versions (whenever feasible) for GMT requirements. The development and adoption of this CBT could be an important way that DoD performs its oversight role.

DoD should perform a gatekeeper role for future GMT requirements. There are many training demands for uniformed DoD personnel, and adding new requirements carries with it both opportunity and financial costs. If new GMT requirements are levied across DoD, the magnitude of GMT requirements can grow, forcing unit commanders to make additional decisions between mission training demands and GMT-mandated requirements. New training requirements that are additive to existing requirements have a big effect and are multiplicative across the force. Although GMT requirements may be valid, they must be balanced with the need to conduct individual specialty and unit collective training, among other requirements. The time impact of new GMT demands must be kept in check and balanced to provide time for mission and specialty training. A single DoD sponsor for GMT should vet new with current training demands to meet the balance and reduce mission-readiness opportunity costs. The goal of this process would be to assess training requirements, increase effectiveness and cost-efficiency, resolve issues, balance priorities, and gain training efficiencies.

The services have indicated that GMT is a burden and can take time away from mission accomplishment. Although our research examines GMT directed by DoD and above, the services also have training topics over which they have direct control. In the appendixes, we identify the extensive topics that the services require. The services have much discretion in reducing or eliminating these training requirements. If the services determine that GMT is a burden, then an intraservice examination of training requirements may indicate that some of these training that could be eliminated, reduced, or standardized.

DoD must engage with and get buy-in from GMT stakeholders. The stakeholders include the service leaders who develop GMT policy and AC and RC service training SMEs. Meetings with them could further assess opportunities to share ideas and consider potential standardization and reduction options. Training SMEs should be continually engaged to illuminate training issues and challenges and to share ideas. Service training programs, although developed independently now, have many similarities. The engagement approach with SMEs should focus on highlighting GMT similarities and reducing training differences.

All existing and new GMT requirements need to be challenged to determine whether they are needed or whether they could be standardized and reduced. DoD should engage with training-topic sponsors to validate training requirements. For example, the DoD EO office is in charge of the DoD EO instruction that requires periodic EO training (DoDD 1020.02). The EO office, among other training-topic sponsors, needs to be engaged to consider required content, periodicity, and reporting requirements for EO training. Involving the key stakeholders will serve to evaluate (and challenge) the training requirements, build consensus for change, and initiate change.

DoD should issue a single DoD directive that lists all GMT requirements. GMT demands are many; the aggregate effect on uniformed service members in terms of time demands across the force is high, especially on the reservists; and guidance that directs that training be completed comes from various DoD directives. A single DoD instruction or directive should list GMT topics and periodicities. This guidance would provide clarity to the services as they evaluate and further develop GMT programs.

Although the focus of this research was on GMT (done by all uniformed personnel), there is a great deal of other training performed that could potentially best be conducted by means of CBT for effectiveness and efficiency. The appendixes list a myriad of training requirements that the services have for their personnel. CBT may perform well in this role in addressing other training demands.

DoD, in conjunction with the services, should review the options that the services have implemented in their GMT curricula. Although the options or approaches taken appear to be having the desired effect of reducing or standardizing training, they have not been analyzed to determine whether they have achieved the desired effects. Moreover, it has not been determined whether there are undesirable consequences of the options. Additional analysis may suggest alternative or additional options to implement.

The Way Forward

The services have made efforts to standardize and reduce GMT. If OSD desires further standardization and reduction of GMT, it should lead this effort. To realize this opportunity, OSD should engage GMT stakeholders in a collaborative fashion to promote the sharing of successes and lessons learned. OSD might stimulate that collaboration by sponsoring a meeting in which the SMEs, responsible for the advances discussed in this report, discuss how they did it and lessons learned that would benefit other services.

Discussions or considerations about the reduction or standardization of GMT requirements should include an examination of the effectiveness or the quality of the training. Training quality must be measured, understood, and maintained at a high level. Further research is needed to examine the quality and effectiveness of GMT. This additional study should evaluate the efficacy of different training delivery methods in terms of how well GMT information is transferred, i.e., what type of information or GMT topics best suit different delivery methods (i.e., stand-up instruction, CBT, or other) and delivery methods that are most effective for the intended audience. This study should evaluate the quality of the content and instructional design to identify additional opportunities to provide standardized training that can be prepared and delivered in less time.

Finally, new training requirements can pose significant time demands across the force. An OSD OPR should be designated to facilitate information sharing that will help the services provide well-designed, service-generic training of all new training requirements, whenever feasible. This could be an important way in which DoD performs its gatekeeper role.

Our Assessment of Topics That Meet the Approved Definition of *General Military Training*

Table A.1
Topics That Meet the Approved Definition of *General Military Training*

GMT Topic	Periodic	Nonoccupational	Directed by DoD or above	Provides Common Knowledge and Skill	Required for All Uniformed Personnel Across All Services	Enhances Ability to Perform Military Duties or Activities	Meets Definition
Advanced traffic safety (Air Force)	N	Y	Y	Y	N	Y	II
Alcohol misuse enabling (Navy)[a]	N	Y	Y	Y	Y	Y	I
Allotments (Marine Corps)	N	N	N	Y	N	Y	III
American Red Cross (Marine Corps)	N	N	N	Y	N	Y	III
Anger management (Navy)	N	Y	N	Y	N	Y	III
Annual ethics training (Air Force)	Y	N	Y	Y	N	Y	II
Army Family Team Building	N	Y	Y	Y	N	Y	II
Army Safety Program	N	N	Y	Y	N	Y	II
Article 137 UCMJ briefing (Air Force)	Y	Y	Y	Y	N	Y	II
Article 137 UCMJ briefing (Marine Corps)	Y	Y	Y	Y	N	Y	II
AT/FP (Army)	Y	Y	Y	Y	Y	Y	II
AT/FP (Navy)	Y	Y	Y	Y	Y	Y	I
BAS (Marine Corps)	N	N	N	Y	N	Y	III
BEPO (Air Force)	N	Y	N	N	N	Y	III
Casualty assistance (Marine Corps)	N	N	N	Y	N	Y	III

Table A.1—Continued

GMT Topic	Periodic	Nonoccupational	Directed by DoD or above	Provides Common Knowledge and Skill	Required for All Uniformed Personnel Across All Services	Enhances Ability to Perform Military Duties or Activities	Meets Definition
CI awareness (Air Force)	Y	Y	Y	Y	Y	Y	I
Club system (Marine Corps)	N	N	N	Y	N	Y	III
CoC (Navy)	Y	Y	Y	Y	Y	Y	I
Combat water survival training (Marine Corps)	Y	Y	N	Y	N	Y	III
Course 1, introductory traffic safety (Air Force)	N	Y	Y	Y	N	Y	II
Course 2, local conditions (Air Force)	N	Y	Y	Y	N	Y	II
Course 3a, intermediate traffic safety (Air Force)	N	Y	Y	Y	N	Y	II
CTIP (Air Force)	Y	Y	Y	Y	Y	Y	I
Defense Eligibility Reporting System (Marine Corps)	N	N	N	Y	N	Y	III
Domestic violence, family violence (Navy)	N	Y	N	Y	N	Y	III
Driver improvement (Marine Corps)	N	Y	Y	Y	N	Y	II
Driver improvement and rehabilitation (Air Force)	N	Y	Y	Y	N	Y	II
Drug abuse prevention and control (Navy)[b]	N	Y	Y	Y	Y	Y	I
Employment and reemployment rights (Army)	N	Y	Y	Y	N	Y	II
EO newcomer's human relations orientation (CONUS) (Air Force)	Y	Y	Y	Y	Y	Y	I
EO: first duty station orientation (Air Force)[c]	N	Y	Y	Y	N	Y	II
EO: human relations (Marine Corps)	Y	Y	Y	Y	Y	Y	I
EO: newcomer's human relations orientation (OCONUS) (Air Force)	Y	Y	Y	Y	Y	Y	I
EO: religious accommodation (Navy)	Y	Y	Y	Y	Y	Y	I

Table A.1—Continued

GMT Topic	Periodic	Nonoccupational	Directed by DoD or above	Provides Common Knowledge and Skill	Required for All Uniformed Personnel Across All Services	Enhances Ability to Perform Military Duties or Activities	Meets Definition
Ethics (Army)	Y	N	Y	Y	N	Y	II
Ethics training (Navy)	Y	N	Y	Y	N	Y	II
Exchange service (Marine Corps)	N	N	N	Y	N	Y	III
Family programs (Marine Corps)	N	N	N	Y	N	Y	III
FHA/VA loans (Marine Corps)	N	N	N	Y	N	Y	III
Food service education (Marine Corps)	N	N	N	Y	N	Y	III
Fraternization (Army)	N	N	N	Y	N	Y	III
Hazing prevention policy (Navy)	N	Y	N	Y	N	Y	III
Health benefits awareness (Army)	N	Y	N	Y	N	Y	III
Health benefits awareness/CHAMPUS (Marine Corps)	N	N	N	Y	N	Y	III
HIV/AIDS (Marine Corps)	N	Y	N	Y	N	Y	III
Homosexual conduct policy (recent legislation repealed this policy) (Army)	Y	Y	Y	Y	Y	Y	I
Homosexuality policy guidance training (recent legislation repealed this policy) (Air Force)[d]	Y	Y	Y	Y	Y	Y	I
IA (Navy)	Y	Y	Y	Y	Y	Y	I
IA awareness training (Air Force)[e]	Y	Y	Y	Y	Y	Y	I
Informal resolution system (Marine Corps)	N	N	N	N	N	Y	III
Information security (Air Force)[e]	N	Y	Y	Y	Y	Y	II
Information security systems (Coast Guard)	N	Y	Y	Y	Y	Y	II
Insurance counseling (Marine Corps)	N	N	N	Y	N	Y	III

Table A.1—Continued

GMT Topic	Periodic	Nonoccupational	Directed by DoD or above	Provides Common Knowledge and Skill	Required for All Uniformed Personnel Across All Services	Enhances Ability to Perform Military Duties or Activities	Meets Definition
Introduction to the Navy's fraternization policy (Navy)	N	Y	N	Y	N	Y	III
Introduction to the Navy's sexual conduct policy (Navy)	Y	Y	Y	Y	Y	Y	I
Introduction to the Navy's tobacco-use cessation policy (Navy)	N	Y	Y	Y	Y	Y	II
Introduction to the stress response continuum and suicide awareness (Navy)	N	Y	Y	Y	Y	Y	I
Lautenberg Amendment briefing (Air Force)	N	N	Y	N	N	Y	II
Law of land warfare (Marine Corps)	N	Y	Y	Y	Y	Y	II
Law of war (Navy)[f]	Y	N	Y	N	N	Y	II
Law of war/detainee operations (Army)	Y	N	Y	N	Y	Y	II
Leadership (Marine Corps)	N	N	N	N	N	Y	III
Legal assistance (Marine Corps)	N	N	N	Y	N	Y	III
Marine Corps Health Promotion Program: Semper Fit (Marine Corps)	Y	Y	Y	Y	Y	Y	I
Marksmanship: pistol (Marine Corps)	Y	Y	N	Y	N	Y	III
Marksmanship: rifle (Marine Corps)	Y	Y	N	Y	N	Y	III
MBST (now Marine Corps Common Skills) (Marine Corps)	N	Y	N	Y	N	Y	III
MCI study/test (Marine Corps)	N	Y	N	Y	N	Y	III
Military justice (Army)	Y	Y	Y	Y	N	Y	II
MWR (Marine Corps)	N	N	N	Y	N	Y	III
NATO security (Air Force)[e]	Y	Y	Y	Y	Y	Y	I

Table A.1—Continued

GMT Topic	Periodic	Nonoccupational	Directed by DoD or above	Provides Common Knowledge and Skill	Required for All Uniformed Personnel Across All Services	Enhances Ability to Perform Military Duties or Activities	Meets Definition
Navy relief (Marine Corps)	N	N	N	Y	N	Y	III
NBC (Army)	Y	Y	N	Y	N	Y	III
NBC defense training (now CBRN training) (Marine Corps)	Y	Y	N	Y	N	Y	III
New-entrant ethics training (Air Force)	Y	N	Y	Y	N	Y	II
No Fear (Coast Guard)	N	N	Y	Y	N	Y	II
No Fear Act (Air Force)	N	N	Y	Y	N	Y	II
Nonnaval medical and dental coverage (Marine Corps)	N	N	N	Y	N	Y	III
Off-duty employment (Marine Corps)	N	N	N	Y	N	Y	III
Operational risk management, time, critical risk management (Navy)	N	Y	Y	Y	Y	Y	II
OPSEC (Air Force)	Y	Y	Y	Y	Y	Y	I
OPSEC: the next generation (Navy)	Y	Y	Y	Y	Y	Y	I
Passenger transportation (Marine Corps)	N	N	N	Y	N	Y	III
Personal financial fitness (Air Force)	N	Y	Y	Y	N	Y	II
Personal financial management (Marine Corps)	N	Y	Y	Y	N	Y	II
Personal financial management (Navy)	N	Y	Y	Y	N	Y	II
Physical fitness (Marine Corps)	Y	Y	Y	Y	Y	Y	I
Physical fitness/ combatives (Army)	Y	Y	Y	Y	Y	Y	I
Physical readiness: preparing for the PRT (Navy)	Y	Y	Y	Y	Y	Y	I
PII (Navy)	N	N	Y	Y	N	Y	II

Table A.1—Continued

GMT Topic	Periodic	Nonoccupational	Directed by DoD or above	Provides Common Knowledge and Skill	Required for All Uniformed Personnel Across All Services	Enhances Ability to Perform Military Duties or Activities	Meets Definition
PR, CoC, SERE (Army)[g]	Y	Y*	Y	Y	Y	Y	I
Prevention of motor-vehicle accidents (Army)	Y	N	Y	Y	N	Y	II
Preventive measures against disease and injury (Army)	N	N	Y	Y	N	Y	II
Privacy Act (Air Force)[e]	N	N	Y	Y	N	Y	II
Privacy Act (Marine Corps)	N	N	Y	Y	N	Y	II
Privacy awareness (Coast Guard)	N	N	Y	Y	N	Y	II
Protection from terrorism, level 1 (Air Force)	Y	Y	Y	Y	Y	Y	I
Public affairs program (Army)	N	Y	Y	Y	N	Y	II
Record management (Air Force)[e]	Y	Y	Y	Y	Y	Y	I
Religious ministries (Marine Corps)	N	N	N	Y	N	Y	III
SAEDA (security awareness) (Army)	Y	Y	Y	Y	Y	Y	I
SAPR (Air Force)	Y	Y	Y	Y	Y	Y	I
SAPR (Army)	Y	Y	Y	Y	Y	Y	I
SAPR (Navy)	Y	Y	Y	Y	Y	Y	I
SAVI (Navy)[i]	N	Y	Y	Y	Y	Y	II
Security education and training awareness (Coast Guard)[h]	Y	Y	Y	Y	Y	Y	I
Sexual harassment (Marine Corps)	Y	Y	Y	Y	Y	Y	I
Sexual harassment/EO/human relations (Army)	Y	Y	Y	Y	Y	Y	I
Sexual health: unplanned-pregnancy prevention (Navy)	N	Y	N	Y	N	Y	III
SGLI (Marine Corps)	N	N	N	Y	N	Y	III

Table A.1—Continued

GMT Topic	Periodic	Nonoccupational	Directed by DoD or above	Provides Common Knowledge and Skill	Required for All Uniformed Personnel Across All Services	Enhances Ability to Perform Military Duties or Activities	Meets Definition
SHAPR (Coast Guard)	Y	Y	Y	Y	Y	Y	I
Space A travel (Marine Corps)	N	N	N	Y	N	Y	III
Sponsorship program (Marine Corps)	N	N	N	Y	N	Y	III
Standards of conduct (CoC) (Marine Corps)	Y	Y	Y	Y	Y	Y	I
Substance abuse (Air Force)	N	Y	Y	Y	Y	Y	I
Substance abuse (Marine Corps)	N	Y	Y	Y	Y	Y	I
Substance abuse and risk reduction (Army)	N	Y	Y	Y	Y	Y	I
Substance abuse-free environment awareness (Coast Guard)	N	Y	Y	Y	Y	Y	I
Suicide awareness (Marine Corps)	N	Y	Y	Y	Y	Y	I
Suicide awareness and prevention (Navy)	N	Y	Y	Y	Y	Y	I
Suicide awareness and violence prevention (Air Force)	N	Y	Y	Y	Y	Y	I
Suicide prevention (Coast Guard)	N	Y	Y	Y	Y	Y	I
Survivor assistance awareness training (Air Force)	N	N	N	N	N	Y	III
Survivor benefit plan (Marine Corps)	N	Y	N	Y	N	Y	III
Team Marine (Marine Corps)	N	Y	N	Y	N	Y	III
TIP basic awareness training (Navy)	Y	Y	Y	Y	Y	Y	I
Training for motorcycle operators on safe riding skills and mishap-avoidance techniques (Air Force)	N	Y	Y	Y	N	Y	II
Transportation of personal property (Marine Corps)	N	N	N	Y	N	Y	III

Table A.1—Continued

GMT Topic	Periodic	Nonoccupational	Directed by DoD or above	Provides Common Knowledge and Skill	Required for All Uniformed Personnel Across All Services	Enhances Ability to Perform Military Duties or Activities	Meets Definition
Troop information (Marine Corps)	N	Y	N	Y	N	Y	III
Voluntary education (Marine Corps)	N	N	N	Y	N	Y	III
Weapon qualification (Army)	Y	Y	N	Y	N	Y	III

NOTE: In the "Meets Definition" column, I = fits the working definition, II = is required by DoD or above but does not meet the working definition for another reason (e.g., periodic, required for all), and III = is required by a DoD component. SAEDA = Subversion and Espionage Directed Against the U.S. Army.

[a] DoDD 1010.4 requires training but does not mandate periodicity.

[b] The Air Force topic is required only for certain occupations.

[c] The DoD policy requires periodic EO training. The Air Force has three EO training topics. This particular training is a one-time event.

[d] The Air Force topic is guidance training and is required only for investigators, commanders, judge advocates, and supervisors.

[e] DoD Regulation 5200 requires annual refresher security training. The Air Force combines the following training under information protection: information security, NATO security, information assurance, record Management, Privacy Act, and Freedom of Information Act.

[f] Training required for detainee operations personnel only.

[g] Only CoC training is required periodically.

[h] The Coast Guard combines AT/FP, information security, and OPSEC.

[i] SAVI combined with SAPR training.

APPENDIX B

Summary of Discussion at the Office of the Secretary of Defense–Hosted Meeting with Service General Military Training Subject-Matter Experts

We met with service SMEs both individually and during an OSD-sponsored GMT summit to share ideas and discuss options to standardize and reduce GMT. A summary of the discussion is provided in this appendix.

In February 2010, SMEs from the Army, Air Force, Navy, and Marine Corps gathered under the direction of OSD to discuss GMT challenges and revamping efforts. At that time, the Air Force had made the most-recent, biggest changes to its ancillary training program. It had reduced time for TFAT to 3.5 hours. To do this, it evaluated whether it was necessary for topics to be conducted annually and then lowered the periodicity for many topics by giving airmen training at different points of their careers, such as accession or schoolhouse attendance. It also reduced training time by evaluating topic time durations and scaled back the time requirements. The other three services were in the process of revising their GMT requirements.

Many issues were discussed during the meeting, such as the number of training requirements, training periodicity, topic audience, the advantages and disadvantages of standardization, delivery methods, and grouping topics. Representatives from different services had varying opinions on many of these topics, but everyone agreed on a few things. First, all participants thought that a single DoD regulation listing all GMT training would be helpful. Currently, all topics are listed in various DoD instructions and directives, and it is hard to keep track of all the requirements. Also, the growth of GMT demands seems to be unchecked, and everyone agreed that a gatekeeper is necessary. Finally, everyone agreed that any attempt to levy training on the forces should be approved through senior leadership.

When Does General Military Training Become a Burden?

The representatives agreed that more discussion on training burdens is necessary. The idea of the GMT burden was discussed, and the definition of *burden* was debated. The questions that were raised included the following:

- At what point is the training requirement too much?
- If the training requirement has become a burden, what is it a burden of: time or resources?
- Whose responsibility is it to answer these questions?
- Whose responsibility is it to resolve these issues?

Training Duration and Periodicity

Another question raised was how much time GMT requirements deserve. On one hand, the services need an adequate amount of time to train. On the other hand, the training duration should not be minimized so much that the services cannot sufficiently address the topic. Perhaps an overarching standard duration is not appropriate.

The suggestion of training differently to different audiences was also discussed. Perhaps three to four hours per year would be sufficient for senior leadership, but more repetition and longer training durations might be necessary for junior enlisted. The Air Force has already incorporated this concept into its continuum-of-learning approach. Junior airmen receive introductory training and later receive awareness training; at leadership levels, individuals receive a higher level of training that teaches them how to deal with the problem at hand, such as substance abuse. The Navy is working on incorporating this concept into its ancillary training as well.

Advantages and Disadvantages of Standardization

A standard GMT project does not exist because each service has its own opinion about what is "right" (because each service's training was developed independently of the other services' training). The motivation for standardization is not to force everyone to conform GMT to a single approach but rather to address training objectives and desired outcomes and to reduce redundancy. To produce these standards, objectives that apply to all personnel must first be determined.

Currently, the only topic that is standardized for all services is IA awareness training. All services must take this DISA-issued training module, which lasts around 60 minutes. The Army has its own supplement to the training that addresses Army-specific IA issues. CTIP is also semistandardized because the Army, Navy, and Marine Corps take the same CBT course, but the Air Force has its own. No other DoD-mandated training topic is standard, and it is up to each service to create its own training modules. Although the services do not want to be told what their training module is, a core curriculum on which the services can build and that they can tailor to their own needs would be helpful.

Time requirements should not be standardized across all services. Some services might choose to reduce training time but increase periodicity. Also, some services have combined similar topics into one training module, which can reduce time.

Training Delivery

Participants in the discussion agreed that training delivery should not be standardized. Some training topics are best delivered in person and are better suited for face-to-face interaction. Examples of these training topics are mostly behavioral topics, such as suicide prevention. Although training for these topics might be most effective in small-group, face-to-face sessions, requiring that they be delivered this way quickly adds to the time burden. Additionally, it is sometimes challenging for units to gather in small groups and find facilitators and meeting spaces. These training sessions are difficult to coordinate and are resource intensive.

Participants opined that CBT is less difficult to conduct and is less resource intensive at the unit end. It allows anyone with access to a computer to complete the training and does not require a trained instructor. Training done through CBT can be completed at the user's convenience. However, CBT does not have the same effect as face-to-face training. Overall, leaders prefer to have options for delivery methods.

Reserve and Guard Perspectives

Many reservists complete their GMT requirements by means of CBT at home on their own time. When members complete these requirements at home, drill weekends can be devoted to mission needs. Sometimes, reservists are given retirement points for their time, but it is rare to be compensated for the hours spent completing ancillary training requirements. The SMEs all agreed that there should be a policy and that money should be available to compensate reservists for their time. It was also discussed that work needs to be done on balancing reservists' heavy GMT requirements with their mission work.

Summary

In summary, service SMEs expressed a desire for tailorable training that could be provided to different audiences. They prefer to have options to provide training through a variety of delivery methods, which would provide flexibility to unit leadership in meeting training demands. SMEs suggested that some like topics could potentially be bundled. In addition, the SMEs suggested that some training be event-driven (e.g., upon reenlistment or PCS). There was discussion of what the "right" periodicity would be for conducting training. Although many training topics are listed as annual, participants wondered whether some periodicities could be extended (or reduced based on content or operations).

There was general agreement among the SMEs that it would be very helpful if one DoD instruction were issued that could consolidate and standardize current GMT guidance.

GMT SMEs expressed the need for tailorable training, a variety of delivery methods, leaving flexibility for delivery of training with unit leadership, consolidating and standardizing DoD guidance, and challenging each aspect of training levied on the force. Additionally, they said, options should be pursued to reduce GMT by examining opportunities to bundle or combine some topics, determine whether training could be event-driven, and determine and challenge the training periodicity.

APPENDIX C

Army, Army Reserve, and Army National Guard Common Military Training

Background

The Army defines CMT in AR 350-1 as consisting of senior leader–selected general training requirements considered essential to individual or unit readiness. This training is common to all members of units, regardless of branch, career field, rank, or grade, and will be incorporated into unit training programs as specified in the governing Army regulation (AR 350-1, Appendix G). The governing guidance for Army CMT is contained in AR 350-1. This regulation provides the required topics and information on certain other topics. GMT requirements are the same for AC and RC. Additionally, Army civilians are responsible for completing the CMT requirements (AR 350-1, pp. 3–17a).

The Deputy Chief of Staff (DCS), G-3/5/7 (Operations, Plans, and Training), biannually validates and publishes a list of general subject areas in which Army personnel require knowledge and skill.

Objectives

The objectives of CMT as defined in AR 350-1 are threefold. First is to limit directed training requirements to the minimum subjects required by law or higher regulatory authority or those validated as essential to individual or unit readiness. Second is to control directed training and ensure that the subjects remain current. This is to provide unit commanders maximum flexibility in achieving and sustaining required levels of proficiency. Third, and more broadly, the principal reason for CMT is to inform the troops about important topics that the military faces. For example, the purpose of the Army's SAPR is to reinforce the Army's commitment to eliminating incidents of sexual assault. Another example is the Army Safety Program, in which the goal is to reduce the risk of death or injury to soldiers and civilians and damage to vehicles, equipment, and property due to accidents.

A subtler objective of CMT is to boost camaraderie and mentorship opportunities between leaders and troops. The Army believes that time spent by seniors with subordinates talking about sensitive issues promotes trust and unit esprit de corps. The Army chooses to utilize time for CMT as an opportunity for a unit leader to mentor the younger troops by delivering as many topics as he or she can through the stand-up delivery method. This both provides a leadership opportunity and reinforces relationships between the leaders and the soldiers.

Curriculum

The Army requires that 21 topics be completed by all service members. AR 350-1 requires training on the following topics:

- Army Family Team Building
- AT
- ethics
- fraternization
- homosexual conduct policy[1]
- law of war and detainee operations
- NBC
- physical fitness and combatives
- PR, CoC, and SERE
- preventive measures against disease and injury
- public affairs program
- SAEDA
- sexual assault
- sexual harassment
- weapon qualification.

The Army CMT includes several Army-specific courses, such as Army Family Team Building and the Army Safety Program.

There are certain CMT topics that are taught in training institutions, which are shown in Table C.1. Training institutions are organized schools that teach standard courses to service members. This is unlike unit-organized and -programmed training, which varies from unit to unit. Therefore, individuals who are en route to their units through the training track receive some common training.

Table C.1
Army Common Military Training Subjects Taught in Training Institutions

Subject	Reference	Proponent
Weapon qualification	AR 350-1/DA PAM 350-38	DCS, G-3/5/7
Physical fitness	AR 350-1/FM 21-20	DCS, G-3/5/7
Combatives	AR 350-1/FM 3-25.150	DCS, G-3/5/7
First aid	AR 40-3	DCS, G-3/5/7
AT	AR 525-13	DCS, G-3/5/7

SOURCE: AR 350-1, Table 3-1.
NOTE: FM = field manual.

[1] A recent revision to law repealed 10 U.S.C. § 654, policy concerning homosexuality in the armed forces.

Guidance

Army regulations require 21 training topics. Army CMT is driven by requirements in law (public law, U.S. Code, executive orders), by DoD directives and instructions, and by Army regulations. For example, CoC training is mandated by an executive order. Executive Order 10631 states,

> All members of the Armed Forces of the United States are expected to measure up to the standards embodied in this Code of Conduct while in combat or in captivity. To ensure achievement of these standards, members of the armed forces liable to capture shall be provided with specific training and instruction designed to better equip them to counter and withstand all enemy efforts against them, and shall be fully instructed as to the behavior and obligations expected of them during combat or captivity. (Eisenhower, 1955)

Fourteen Army training topics are mandated by DoD. Army regulations drive other required CMT, including weapon qualification, NBC, public affairs program, and fraternization. AR 350-1 is currently being rewritten and evaluated.

Delivery

The Army employs several delivery methods, such as stand-up delivery, CBI (interactive and noninteractive), and videos. *Stand-up delivery* is synonymous with classroom instruction or a command presentation. This type of presentation involves a leader—typically an NCO training officer—speaking in front of a classroom of soldiers. Usually, but not always, the presentation will be accompanied by PowerPoint slides that guide the speaker. Sometimes, if the commander sees it as beneficial, stand-up delivery will be interactive and will include a discussion during or after the lecture. Attendance is taken manually by a muster sheet, which is the tracking method of completion for stand-up delivery.

CBT is a common delivery method for several CMT topics. CBT can be categorized by two methods: interactive and noninteractive. A CBT course is considered to be interactive if it requires some interaction from the user besides "clicking through" slides. A test, pre/post or during the CBT course, would qualify it to be interactive. A noninteractive CBT course would allow the user to simply click through the training slides and still qualify as completed. To deter individuals from clicking through the training as quickly as possible, Army CBT courses often require soldiers to spend a certain amount of time on each screen before proceeding to the next slide. A printed certificate is proof of completion for CBT courses.

Many of the Army's CMT topics are required to be delivered hands on. Weapon qualification, physical fitness, and NBC defense training are all examples of hands-on training. Commanders or senior military supervisors establish physical fitness training programs, and each soldier must meet the physical fitness standards as measured during the Army Physical Fitness Test (APFT) (AR 350-1, pp. 1–24c). Every soldier must meet weapon qualification standards laid out by U.S. Department of the Army Pamphlet (DA PAM) 350-38 and be proficient in the safe use, maintenance, and employment of his or her weapon (DA PAM 350-38). NBC training culminates with soldiers entering and exiting the gas chamber.

No single delivery method is mandated for Army CMT topics, and multiple methods are usually available. Cultural and resource differences between AC and RC lead to different preferred delivery methods. Also, each unit or command is in charge of determining which delivery method to utilize for each topic. This flexibility in choices leads to various delivery methods for each topic and, often, various versions of the same delivery method.

The Army's AC has the benefit of having its soldiers "365 days per year" and therefore has more flexibility as to how it deliver its training and how long it takes to do so. In a stand-up delivery setting, a unit commander will decide how long the training will take. For example, he or she can choose to complete the training in 30 minutes or decide that the troops would get more out of an interactive discussion that takes two hours.

Army CMT SMEs indicated that stand-up delivery is the preferred choice for the Army's AC because it allows an opportunity for the leader to talk directly to his or her soldiers. The more-sensitive topics are viewed as more effective when delivered face to face. For example, a delicate topic, such as suicide prevention, is probably best delivered face to face, but a less thorny topic, such as AT, might be just as effective if conducted via CBT.

RC SMEs indicated that the RC attempts to complete the CMT topics through stand-up delivery during drill periods because not everyone has access to a computer. Army RC officials indicated that reservists who miss training during drill periods are instructed to complete that training on their own time via CBT. Each unit is in charge of keeping track of its soldiers' training completion rate and informing each soldier how to complete the training online.

The Army National Guard approach to completing CMT is that members are encouraged to do all the available training online via CBT so that drill periods are not filled with CMT. Although the members often complete this training on their own time, they are not compensated for the time used to do the training. Before distance learning was available, an entire weekend would be devoted to CMT topics.

Tracking

Within the Army AC, each commander is responsible for keeping track of his or her soldiers' training completion. The commander provides a unit status report to his or her superior that contains statistics on that unit's training completion. This report is the unit commander's motivation to have every soldier complete training on time. The Army's aim is to have all training completion recorded in the Digital Training Management System (DTMS).

In the Army Reserve, an NCO training officer is responsible for preparing CMT, verifying attendance, and tracking the unit's training completion. Each unit has a training NCO to carry out this responsibility, and the inspector general is considered to be the training watchdog. The Army National Guard's training is tracked using the Digital Training Management System, which is used to record and track individual and unit training.

Periodicity

The periodicity of Army CMT is as directed by AR 350-1. There are four Army training topics that are required annually:

- AT/FP[2]
- ethics[3]
- law of war/detainee operations[4]
- physical fitness/combatives.[5]

AR 600-20, pp. 6–15, mandates that sexual harassment training be conducted quarterly:

> Leaders will conduct mandatory unit EO/prevention of sexual harassment training quarterly. At a minimum, two of the quarters will consist of Prevention of Sexual Harassment training. The other two quarters will consist of training that is interactive, small group, and discussion-based.

Most Army topics do not have any mandated periodicity requirements.

Summary

Table C.2 provides a summary of Army CMT requirements. In the first column are the 25 topics mandated by AR 350-1. The reference directing the training is contained in the second column. These references include DoD, public laws, U.S. Code, executive orders, or an Army regulation that mandates the topic. Army regulations mandate training in several topics. The periodicity of each topic is designated in the AR column and is explained in the notes.

The guidance that directs the Army's CMT varies by topic. There are ten training topics mandated by higher authority, such as U.S. Code, public law, or executive order:

- Army Safe Program
- employment and reemployment rights
- ethics
- homosexual conduct policy
- military justice
- PR, CoC, and SERE
- preventive measures against disease and injury
- SAEDA
- SAPR.

---

2 DoD AT standards require providing "post-accession Level I AT Awareness Training annually to all DoD personnel." See DoDI 2000.16.

3 Office of Government Ethics, 1990, Subpart G, states,

> Each agency must have an ethics training program to teach employees about ethics laws and rules and to tell them where to go for ethics advice. The training program must include, at least, an initial agency ethics orientation for all employees and annual ethics training for covered employees.

4 AR 350-1, pp. 4–18c states that training is conducted annually and conducted again prior to deployment when directed by a deployment order or appropriate authority.

5 DoDI 1308.3, § 6.1.3.2 states, "All Service members, regardless of age, will be formally evaluated and tested for record at least annually unless under medical waiver."

Table C.2
Army Common Military Training Requirements in Units

Subject	Reference	Periodicity
Army physical fitness training	DoDD 1308.1	O
Army Substance Abuse Program	DoDD 1010.4	I, A, R
Army Suicide Prevention Program	DoDD 6200.04	A, P, R
Army Traffic Safety Training Program	DoDI 6055.04	I
Army Warrior Training	AR 350-1	A
AT training	DoDI 2000.16	A
CBRN defense training	AR 350-1	A
Composite risk management	AR 385-10	O
CTIP program	DoDI 2200.01	A, P
Cultural awareness training	AR 350-1 Chapter 8	P
Employment and reemployment rights (RC only)	28 U.S.C. §§ 4301–4335	A, P, R
EO program	DoDD 1350.2	S
Ethics	18 U.S.C. § 202(a)	Initial entry
Fraternization	AR 600-20	A
Homosexual conduct policy	10 U.S.C. § 654	A
Law of war/detainee operations	DoDD 2311.01E	A, P
Modern Army Combatives Program	AR 350-1	O
OPSEC	DoDD 5205.02	I, A, P, R
PR	Executive Order 10631	A
Prevention of sexual harassment	DoDD 1350.2	S
Preventive measures against disease and injury	Executive Order 12196	P
Resilience training	AR 350-1	P, R
SAEDA	NSC-NSDD-197	A
SAPR program	Pub. L. 108-375 §577	A, P, R
Weapon qualification	DA PAM 350-38	S (AC), A (RC)

NOTE: NSC-NSDD = National Security Council national security decision directive. The periodicity codes are as follows: A = annual (trained annually). I = in-processing (required for an individual upon assignment to a new unit). O = ongoing (continuous training, not a single event). P = predeployment (addressed before the unit is deployed on an operational mission). R = redeployment (addressed upon redeployment from an operational mission). S = semiannual (trained twice per year).

Fourteen of the Army's 25 topics are mandated by DoD either by directive or instruction. There are six topics that are mandated by the Army only and not by either DoD or a higher authority:

- Army Warrior Training
- CBRN defense training
- composite risk management
- cultural awareness training
- fraternization
- Modern Army Combatives Program.

Reserve Component and Guard Perspectives

Reservists meet for one weekend per month and two weeks out of the year, so their time devoted to the Army is limited. However, the reservists are required to do the same training as their AC counterpart, which results in too much training time required for the time allotted for the reservists.

Our discussion with ARNG and Army Reserve SMEs indicated that many Army RC units have their members complete the CMT requirements on their own time at home. The SMEs also indicated that the RC has to keep retention as a priority, and having an entire drill period devoted to CMT topics is not good for retention. Reservists and ARNG members have civilian jobs and often sign up for the military to support the mission and to be a soldier. The RC SMEs whom we interviewed noted anecdotally having dedicated an entire RC drill day to PowerPoint briefings, e.g., EO and substance abuse prevention training, which was not popular with the troops and does not help promote retention. The SMEs stated that soldiers would rather perform their jobs (as opposed to sitting through lectures or presentations) that are more directly related to their specialty and that support mission readiness.

Under U.S. Code, it is legal for reservists to be compensated for training that they complete at home. Currently, reservists and ARNG members are not compensated for the training that they complete on their own time, but the Army is looking into how compensation could be provided. Thus, it is necessary to have good front-line leaders who make sure their troops get the CMT done without compensation.

Navy and Navy Reserve General Military Training

Background

The Navy requires that "all hands" receive periodic training on GMT topics. The Navy defines *GMT* as nonoccupational general training required periodically for all Navy personnel and taught at the command level. The governing guidance for Navy GMT is contained in OPNAVINST 1500.22F and a recent fleetwide message (NAVADMIN 098/10). The updated requirements for periodic GMT are contained in this message. GMT requirements are the same for both Navy AC and RC military personnel, i.e., all Navy uniformed personnel must complete the prescribed GMT.

The Navy considers GMT to be an important mechanism for emphasizing leadership responsibilities and core values for both officers and enlisted personnel.

Objectives

The objective of Navy GMT is to inform and motivate sailors on both personal and professional levels related to their naval careers. GMT topics fall into several categories, including personal growth and professional relationships, risk management, and wellness. These training areas and topics serve to reinforce the professional attitudes and behaviors of all sailors to enable them to become successful members of the Navy's combat team.

Curriculum

Over the past several years, uniformed Navy personnel have received training in 26 GMT topics. Navy-wide GMT training topics have been directed each fiscal year to Navy commands, normally through a fleetwide message.[1] The FY 2009 GMT message lists 13 training topics that uniformed Navy personnel must complete.[2] However, in FY 2010, the Navy significantly changed GMT training requirements. The FY 2010 guidance now requires only five GMT topics to be done by all hands in FY 2010 and FY 2011 (NAVADMIN 098-10):

- responsible use of alcohol (use, prevention, and control)
- responsible personal behavior (SAPR, EO and sexual harassment grievance procedures)

[1] The Navy GMT instruction, OPNAVINST 1500.22F, does not list the GMT topics.

[2] For example, FY 2009 GMT requirements were provided by NAVADMIN 033/09.

- ask-care-treat (suicide awareness and prevention)
- improving personal financial management
- operational stress control.

The Navy removed three topics from annual GMT but requires these topics to be done upon a sailor's reenlistment:

- hazing policy and prevention
- fraternization awareness and prevention
- homosexuality policy.

Additionally, Family Readiness Program training will be required for reenlistees in FY 2011.

Guidance

Navy GMT is driven by requirements in law (e.g., Public Law and U.S. Code), by DoD directives and instructions, and by Navy policies contained in service directives and instructions.

Delivery

The Navy changed the policy and procedures for delivery of GMT. The Navy now considers the command-scheduled presentation to be most effective if persons in authority show an active interest by attending the training or being involved in the instruction.

In the past, the options to complete training were through Navy e-learning (online training or via CD-ROM) or as a command presentation (instructor-guided training). Now, per recent guidance (NAVADMIN 098-10), GMT topics will be delivered via instructor-led training sessions provided at the unit by command leadership, command training teams, and collateral duty officers or chief petty officers.

The Center for Personal and Professional Development (CPPD) is responsible for maintaining and updating library topics. Facilitator guides and presentations are used by Navy trainers and downloaded from NKO (U.S. Department of the Navy, undated). The downloadable material supports the standardization of training throughout the fleet. Presentation materials include PowerPoint slides and videos that the training facilitator shows, reviews, and discusses with the trainees.

Tracking

Unit training officials document GMT completion by entering GMT into learning-event completion forms. The documentation is centrally tracked via Fleet Training Management and Planning System (FLTMPS).

Periodicity

Navy officials provided the research team with a list of GMT topics completed by uniformed personnel for the past several years, specifically for FY 2006 through FY 2009. The topics varied year to year. Generally, commands were required to administer 12 topics per year; however, in FY 2009, 13 topics were required. As noted earlier, the Navy's GMT guidance for FY 2010 and FY 2011 reduced the number of GMT topics to five.

As also noted earlier, the Navy removed hazing policy and prevention, fraternization awareness and prevention, and homosexuality policy GMT topics from the annual GMT listing. However, these topics are now required training for sailors reenlisting in FY 2010. The training materials for reenlistment training are available via the NKO.

Summary

Of the 26 Navy GMT requirements imposed over the past several years, 21 are directed by DoD guidance. DoD guidance varies in specificity regarding how often the training must be completed. Some directives direct the services merely to develop training plans, while others direct the services to conduct annual training. Five DoD-directed Navy GMT requirements must be completed annually:

- AT/FP
- ethics training[3]
- law of war
- PII
- TIP basic awareness training.

The Navy has reduced many mandatory GMT requirements and now "highly encourages" commanding officers to establish recurring refresher training in selected topics:

- anger management
- CoC level A
- drug abuse prevention and control
- operational risk management
- OPSEC
- physical readiness
- tobacco use prevention and cessation.

Changing GMT from required to "highly encouraged" gives the commanding officer the discretion about GMT. Unit commanding officers are now responsible to decide what GMT refresher training is needed and the periodicity.

The Navy intends to target or tailor GMT to specific audiences (nonsupervisory, supervisory, and command leaders) in the future. The merits of instructor- or leader-led classroom

---

[3] The DoD directive on ethics training implements higher-level ethics required training, required for senior-level personnel.

training, e.g., providing greater opportunities for leadership and mentorship than CBT, are also being factored into future plans for training.

Reserve Component Perspectives

The challenges experienced by Navy RC personnel in completing GMT requirements that were presented to us fall into three major areas: time, resources, and compensation. Navy Reserve training officials also offered views on the way ahead for GMT.

Time is a scarce and valuable resource for Navy reservists. Reservists normally drill one weekend per month and two weeks per year. RC personnel must balance their time to provide mission support while also meeting administrative, occupational, and general training demands. Reserve officials indicated that a Navy GMT topic is normally done once per month and completed via NKO. Although most topics are delivered through NKO, some are given via stand-up training. AC and RC personnel sometimes complete the training together, when colocated. All RC personnel are supposed to complete the same training every month.

Completing GMT is more challenging now than in the past for Navy reservists. During the 1990s, there were greater numbers of Navy RC personnel, and more people were available to complete mission-readiness work than at the present. Today, Navy SMEs note that there are far fewer reservists who are doing much more work (providing mission support). The combination of fewer reservists and mandatory GMT requirements can stress the time available to provide the valuable mission support required.

Reserve officials voiced concerns that Navy RC personnel do not have the same access to resources that AC personnel use to complete GMT. The access relates to availability of computer resources and varies by occupation. For example, in a reserve aviation squadron maintenance detachment, there is a limited number of computers for drilling reservists to use during a drill weekend. AC personnel, who are on duty 24/7, have time and access to computers to complete a GMT course via NKO. During a drill weekend with a large number of drilling reservists on duty, RC personnel have limited access to a limited number of computers. When Navy RC personnel are providing maintenance or mission support to the squadron, as well as conducting or receiving occupational training, a lack of resources combined with competing demands on their limited time to provide mission support results in challenges in completing individual GMT requirements.

The Navy Reserve does have additional funding available for Navy reservists to perform additional drills; however, officials indicate that this funding is generally reserved for senior Reserve Center personnel who put in much additional time in planning and administering the reserve unit's activities. Although Navy RC personnel can complete GMT on their own time at home, GMT is not normally done outside the drill periods. Navy officials indicated that GMT requirements can be completed in unpaid status, i.e., the reservists can earn retirement points for completing the training, but they will not receive drill pay for the time spent completing GMT.

Navy Reserve commands have and desire flexibility in choosing the delivery methods used to complete GMT. Generally, in the Navy Reserve, GMT is completed via NKO; however, GMT is sometimes provided via a CD-ROM or classroom.

APPENDIX E

Air Force, Air Force Reserve, and Air National Guard Ancillary Training

Background

The Air Force has made extensive changes to its ancillary training program. There were increasing mission demands on airmen's time, and the Air Force decided that it was necessary to balance training demands and tailor training to the audience in need of the training. The training must also be done at a periodicity that ensures retention, and it should be no longer than what is needed to provide the information.

The Air Force defines *ancillary training* as universal training, guidance, or instruction, regardless of AFSC, that contributes to mission accomplishment. It does not include functional, occupational, or additional duty training (AFI 36-2201). The governing guidance for Air Force CMT is contained in AFI 36-2201, Vol. 1.

Objectives

The Air Force ancillary training is divided into four categories:

- annual TFAT
- selected force training
- event-driven training
- EST.

TFAT is general awareness-level training for the Total Force, which is given via CBT.[1] Selected force training is ancillary and is targeted to specific audiences, e.g., survivor assistance training for commanders. Event-driven ancillary training is that which is required to be done during certain events, e.g., required training when reporting to a new duty station. EST is directly related to an airman's ability to survive and operate in a contingency environment. Overall, the Air Force objectives in regard to ancillary training are to streamline and tailor the training to meet the specific needs of the airman and the audience.

[1] Air Force SMEs stated that the method of instruction for one annual TFAT topic, Sexual Assault Prevention and Response (SAPR), is given via stand-up lecture.

Curriculum

Annual TFAT is required for all Air Force personnel. The training is done via CBT modules via the Air Force ADLS website. Table E.1 lists annual TFAT, a description of the training, and the training block or subject area. Training blocks provide a group of CBT topics in one module.

There are five additional selected force training requirements that are provided either by local training experts or via CBT on ADLS (see Table E.2). This training is tailored to specific audiences and is done annually (except for homosexuality policy guidance training for supervisors upon assignment to a supervisory role).

In addition to the TFAT and selected force ancillary training, the Air Force has event-driven training that is triggered by an event, such as PCS in-processing or arrival at first duty station. There are 14 event-driven ancillary training topics, which are either one-time

Table E.1
Air Force Annual Total Force Awareness Training

Course	Description	Block or Subject Area
CTIP	Educates personnel on the worldwide trafficking menace, national TIP policy, and attendant responsibilities	Human relations
IA awareness	Ensures that personnel are aware of the latest threats to computer security issues and how to protect against them	Provided on separate module on ADLS
Information security	Ensures security and protection of DoD information	Information protection
NATO security	Covers NATO-classified information and rules for handling	Information protection
OPSEC	Ensures that OPSEC is integrated into all elements of job performance	Information protection
Privacy Act training (with FOIA)	Ensures that PII is protected when designing, developing, operating, or maintaining any system of records	Information protection
Protection from terrorism, level 1	Ensures that personnel are vigilant about possible terrorist actions and can employ AT tactics, techniques, and procedures	FP
Record management	Ensures that personnel are aware of legal requirements for management of records	Information protection
SAPR	Educates Air Force members on the cycle of reporting, response, and accountability, including the role of installation sexual assault response coordinators	Delivered via live presentation format
Suicide awareness and violence prevention	Raises awareness of issues surrounding suicide and violence and provides techniques to help airmen prevent these incidents	Human relations

Table E.2
Air Force Selected Force Ancillary Training

Course	Description	Frequency/Audience
Annual ethics training	Enhances understanding of federal ethics laws and regulations	Annually for filers of financial disclosure reports, presidential appointees, other employees designated based on their official duties, and contracting officers
Homosexuality policy guidance training	Trains leadership on proper handling of homosexuality issues	Annually for investigators, commanders, and judge advocates; one time for supervisors (raters of military personnel) upon assignment to supervisory role
Lautenberg Amendment briefing	Commanders' annual briefing and certification regarding limitations on carrying firearms for members convicted of certain misdemeanors or domestic violence	Annually for all Air Force personnel who regularly handle firearms or ammunition, upon PCS or permanent change of assignment or when temporary duty assignments will require use of firearms or ammunition
No Fear Act	Enhances employee understanding of federal whistleblower protections	Biannually for federal civilian employees and their supervisors for awareness of antidiscrimination and whistleblower protection laws
Survivor assistance awareness training	Trains Air Force members who would be directly involved in assisting survivors in the aftermath of a death	Annually for commanders, first sergeants, and selected unit members

events (such as personal financial fitness) or recurring. Table E.3 lists the event-driven courses, descriptions, and the events that trigger the training requirement. Unit or base personnel provide this training.

Guidance

The Air Force took action and made several changes in its approach to ancillary training in 2007. It convened a working group in Air Force Smart Operations for the 21st Century (AFSO21) to review and reduce the ancillary training requirements. Our discussions with SMEs indicate that there was a push to do more training via CBT. A guidance memorandum from Headquarters U.S. Air Force Manpower and Personnel (HQ USAF/A1) promulgated ancillary training changes to AFI 36-2201 (AFI 36-2201, Vol. 1). This guidance divided the training into the four categories addressed above (annual TFAT, selected force, event-driven, and EST) and eliminated 11 training topics. The guidance further stated that ADLS (online training via CBT) is the preferred method for airmen completing TFAT. Training may also be downloaded and can be presented by an instructor using the ADLS presentation. A benefit of completing training via ADLS is that it automatically updates an airman's training accomplishments in his or her training jacket, an online tool that keeps track of airmen's training completion status. The Air Force continued to refine the training, and the most-recent guidance, AFI 36-2201, Ancillary and Expeditionary Skills Training Policy (AFI 36-2201, Vol. 1), was issued November 17, 2009.

Table E.3
Air Force Event-Driven Ancillary Training

Course	Description	Event Trigger
Advanced traffic safety	Training on consequences of bad driving behavior	Provided approximately one year after arrival at first duty station
Article 137 UCMJ briefing	Training on the UCMJ	(1) Initial entry on active or reserve duty; (2) 6 months of active duty or completion of basic training for reservists; (3) upon reenlistment
BEPO	Awareness training to assist in preparing for emergency situations; location specific	PCS in-processing
CI awareness	Training to prevent espionage, sabotage, and treason	PCS in-processing
Driver improvement and rehabilitation	Training on safe driving techniques; encourages positive driving decisionmaking and teaches the consequences of poor decisions	Commander- or supervisor-directed upon receipt of certain traffic violations or incidents
EO first duty station orientation	EO and human relations awareness training for all new personnel	First duty station, initial unit
EO newcomer's human relation orientation (CONUS)	Training on DoD, Air Force, and installation commander or center director EO policies and local human relations climate issues	PCS in-processing
EO newcomer's human relation orientation (OCONUS)	Training on DoD, Air Force, and installation commander EO policies, local human relations climate issues, and host-nation customs and courtesies	PCS in-processing
Intermediate traffic safety	Training on the top ten causes and countermeasures to prevent traffic mishaps	First duty station, initial unit
Local conditions	Training to enhance individual safety awareness when assigned to a new location	PCS in-processing
New-entrant ethics training	Ethics training for all new personnel	New entrant: Within 90 days of beginning work for the Air Force, first duty station, initial unit
Personal financial fitness	Financial management education for all new airmen	First duty station, initial unit
Substance abuse education	Training on substance abuse recognition and prevention techniques	First duty station, initial unit, and PCS in-processing
Training for motorcycle operators (experienced and basic)	Training for new and experienced motorcycle operators on safe riding skills and mishap-avoidance techniques	One-time training required to operate a motorcycle

An important outcome of the Air Force examination and decisions about ancillary training is that it has developed a gatekeeper process. This process examines new training requirements, vets them through the Air Force Learning Center (AFLC), to advise leadership with the full awareness of impact of new training requirements so that leaders can establish priorities and set limits.

Delivery

The Air Force employs several delivery methods, such as stand-up delivery, CBT (interactive and noninteractive), and videos. As noted earlier, the preferred method for completing TFAT is via CBT on the ADLS website. One of the goals of the Air Force examination of TFAT requirements was to deliver the training in three 30-minute training modules. SAPR training was exempted from this requirement because the Air Force mandated that it be conducted in a classroom or small-group environment because of the sensitivity of the topic. Our discussions with Air Force officials indicate that training via CBT may not be as effective as instructor-led training for some non-TFAT work, e.g., SAPR.

Tracking

When an airman completes ancillary training via ADLS, his or her electronic training folder, which indicates the status of completion of required training, is automatically updated. However, when training is provided via an instructor-led session, training is tracked via muster sheets and entered into the individual's and the unit's electronic records manually.

Periodicity

We noted earlier that there are ten TFAT courses that must be completed annually by all Air Force personnel. There are five that must be done by selected forces annually. In addition, there are 14 event-driven training requirements, and the periodicity of event-driven training varies.

Summary

The Air Force has made progress in reducing the number of training topics and reducing the length of training. It has also provided direction on the preferred method to complete TFAT, i.e., online via ADLS. Completing training online reduces the secondary requirement to record and track training because it is recorded electronically. There is little testing with the Air Force TFAT online approach to training. The Air Force gatekeeper process, which challenges the need for new training requirements and assesses their impacts, can prevent training load creep.

Reserve Component and Guard Perspectives

Like other reservists, AFR and ANG personnel meet for one weekend per month and two weeks out of the year, so the time that they can devote to the Air Force is limited. However, the reservists are required to do the same training as their active counterparts, which results in too much time required for training for the total time allotted for the reservists.

Air Force Reserve

The reserves mirror the AC for ancillary training requirements completely. AFR SMEs stated that the 39 days they have for training are completely full and that the additional man-days they are authorized are also full of mission demands and training requirements.

For training delivery, AFR desires flexible delivery methods. It needs the availability of CBT and the ability to perform mass briefings for all topics. The resource challenges that it faces include the cost of training, compensation for training, and limited facilities (e.g., computers, classrooms) needed for training. AFR indicated that maintenance units might not have a computer for each reservist and that mass briefings (or stand-up lectures) would be beneficial for these units.

The units focus on mission support during weekend drill periods, and ancillary training competes for time with skill training. Overall, the AFR officials indicated that there are stresses on the units' and the individuals' time, that the burden on citizen airmen is growing, and that resources (money for extra drills) have to keep up with increased commitments.

AFR officials expressed the need for the availability of CBT and that they would like the opportunity for mass briefings to be offered for all topics that are required to be delivered via stand-up delivery. Most, but not all, topics are available via CBT.

SAPR training is a challenge because the Air Force mandates that it be delivered in small groups. The SAPR small-group delivery is a strain on reserve resources because the subject is challenging to cover in small groups with limited reserve drill time available. AFR officials added that, although suicide prevention is available via CBT, many units also do this training in a stand-up presentation format.

The best option for delivery methods from the AFR officials' view is by using a buffet of delivery methods. The commander's discretion should determine the delivery method. Most use a combination of training delivery methods now. CBT is available via ADLS, CDs, and the ARCNet website. SMEs use the CBT (downloaded) format to lead a large group; usually, this is accompanied by a discussion. For someone who missed the training, CBT is a good option. Small-group training can be done, but it puts a strain on resources.

Air National Guard

Most ancillary training instructors are ANG personnel, but SAPR SMEs come to the command to provide that training. They use ARCNet (a website for duty planning, scheduling and tracking, readiness training, budget management, and self-inspection training) as the platform to manage and track training.

The ANG tries to do training en masse because not all ANG personnel have access to computers, e.g., aircraft maintenance personnel. En masse TFAT can be downloaded, but it is the same training that an individual would do. Normally, an SME will teach the course to ANG personnel. However, there is no extra pool of instructors to lead the training. Preparing for and leading or providing the training take time out of someone's training schedule.

ANG representatives indicated that they prefer to have multiple formats to deliver training.

SAPR training is one exception to en masse training because it must be delivered in small groups of four to five personnel. This training approach puts a big strain on resources when the reserve wing is large. OPSEC training must be taken individually in ADLS.

Beyond DoD-mandated training, some states sometimes mandate more training on top of what the Air Force requires, e.g., hurricane training or wildfire fighting. It is challenging for ANG personnel to meet training demands because they must conform to both the Air Force requirements and those of their respective states.

ANG SMEs voiced concern that there is too much training to be done with the resources available. Every unit adjusts how it conducts its ancillary training. In some units, it is conducted once a month; in others, a full weekend of ancillary training is offered once per year. The training approach is at the discretion of the wing commander.

Some states allow members to telecommute to complete training, and then it is left up to the unit commander to decide whether to allow telecommuting.

ANG officials indicated that there are many pulls on ANG members' time when they drill. During drill weekends, flying missions must be accomplished. Meeting these many mission demands requires flight crews, maintainers, and administrative staff. ANG units are made up of airmen at varying skill and experience levels, and all personnel must complete ancillary training. ANG time demands are also stressed by the need for physical training, medical and dental physical exams, administrative demands, and record reviews, as well as medical training demands (self-aid and buddy care). Overall, the ANG is challenged to meet the missions and ancillary (and other) training requirements. SMEs stated that the ANG is asking for "too much sugar for a nickel."

APPENDIX F

Marine Corps and Marine Corps Reserve Ancillary Training

Background

The Marine Corps defines *ancillary training* as "training which is directive in nature, but which should be prioritized below mission-related and formal training requirements when planning" (MCRP 3-0A). The overarching directive that consolidates and communicates ancillary training requirements, "Unit Training Management Guide," MCRP 3-0A, was last issued in 1996. Since then, the Marine Corps ancillary training program has changed significantly. The Marine Corps is currently in the process of updating its overarching directive to incorporate the changes that have occurred. Discussions with Marine Corps training personnel indicated that the Marine Corps and Marine Corps Reserve consider ancillary training to be an opportunity for commanders to exercise leadership and mentorship with their subordinates.

Objectives

The Marine Corps's ancillary training program is integrated into its overall training philosophy and principles. These principles are (1) train as you fight, (2) make commanders responsible for training, (3) use standards-based training, (4) use performance-oriented training, (5) use mission-oriented training, (6) train the Marine Air Ground Task Force to fight as a combined arms team, (7) train to sustain proficiency, and (8) train to challenge (MCRP 3-0A). These principles further emphasize the unit commander's role in being personally involved in all levels of training. These principles reflect the integrated way in which the Marine Corps approaches training its personnel.

Curriculum

The Marine Corps requires 48 topics that must be completed by all, or nearly all, marines.[1] It divides these into two categories: formal and ancillary. Formal training requirements are defined as those that "are established by Marine Corps orders and directives" and those that define the Marine Corps as a service. Ancillary training requirements "are additional, secondary training requirements" that provide information to improve the administrative, social,

[1] Depending on assignment or past qualifications, some marines are exempt from training, such as rifle or pistol marksmanship and combat water survival. See MCO 3574.2K and MCO 1500.52C.

or cultural aspects of military service (MCRP 3-0A). Lists of formal and ancillary training requirements are presented in Tables F.1 and F.2, respectively.

There are seven topics named by the Marine Corps in Appendix C of MCRP 3-0A that also fall within the RAND definition of *GMT*:

- physical fitness
- sexual harassment
- substance abuse
- Privacy Act
- CoC
- suicide awareness
- EO and human relations.

The Marine Corps physical fitness program is defined in MCO 6100.13. One of its objectives is to "Contribute to the overall health and wellness of every Marine through regular exercise, proper nutrition, health education and periodic physical and combat fitness evaluations." The program includes both conditioning programs and annual tests for both AC and RC.

The Marine Corps sexual harassment training program is required in MCO 1000.9A. Its intent is to ensure that "All Marine Corps personnel will treat each other with dignity

Table F.1
Formal Training Requirements for the Marine Corps

Topic	Highest Reference	Periodicity
Combat water survival training	MCO 1500.52C	M
Law of land warfare	DoDD 2311.01E	A
Leadership	MCRP 3-0A	A
Marksmanship, pistol	MCO 3574.2K	A
Marksmanship, rifle (BCRM, ICRM, ACRM)	MCO 3574.2K	A
MBST (now Marine Corps Common Skills)	MCO 1510.121A	A
MCI study/test	MCRP 3-0A	A
NBC defense training (now CBRN training)	MCO 3400.3F	A, O
Physical fitness	DoDD 1308.1	O, S
Sexual harassment	DoDD 1350.2	A
Substance abuse	DoDD 1010.4	A
Troop information	MCRP 3-0A	A

NOTE: BCRM = basic combat rifle marksmanship. ICRM = intermediate combat rifle marksmanship. ACRM = advanced combat rifle marksmanship. In the "Periodicity" column, A = annual (trained annually); M = multiple periodicities apply, depending on the marine's previous level of qualification or, in the case of troop information, at the commander's discretion; O = ongoing (continuous training, not a single event); S = semiannual (trained twice per year). Troop information is estimated to require approximately eight hours annually and is intended to cover a rotating selection of topics from the list of ancillary training requirements. MBST = Marine Battle Skills Training.

Table F.2
Ancillary Training Requirements for the Marine Corps

Topic	Highest Reference	Periodicity
Allotments	MCRP 3-0A	M
American Red Cross	MCRP 3-0A	M
Article 137 UCMJ	UCMJ Article 137	M
BAS	MCRP 3-0A	M
Casualty assistance	MCRP 3-0A	M
Club system	MCRP 3-0A	M
CoC	Executive Order 10631	A
DEERS	MCRP 3-0A	M
Driver improvement	DoDI 6055.04	M
EO/human relations	DoDD 1350.2	A
Exchange service	MCRP 3-0A	M
Family programs	MCRP 3-0A	M
FHA- and VA-insured loans	MCRP 3-0A	M
Food service patron education	MCRP 3-0A	M
Health care benefits/CHAMPUS	MCRP 3-0A	M
HIV/AIDS	MCRP 3-0A	M
Informal resolution system	MCRP 3-0A	M
Insurance counseling	MCRP 3-0A	M
Legal assistance	MCRP 3-0A	M
Marine Corps health promotion program, Semper Fit	DoDD 1308.1	A, O
MWR	MCRP 3-0A	M
Navy relief	MCRP 3-0A	M
Nonnaval medical and dental coverage	MCRP 3-0A	M
Off-duty employment	MCRP 3-0A	M
Passenger transportation	MCO 4600.40A	M
Personal financial management	SECNAV 1740.4	M
Privacy Act	5 U.S.C. § 552a	A
Religious ministries	MCRP 3-0A	M
SGLI	MCRP 3-0A	M
Space A travel	MCRP 3-0A	M
Sponsorship program	MCRP 3-0A	M
Suicide awareness	DoDD 6200.04	A

Table F.2—Continued

Topic	Highest Reference	Periodicity
Survivor benefit plan (part of survivor annuity programs)	MCO 1741.11C	M
Team Marine	MCRP 3-0A	M
Transportation of personal property	MCRP 3-0A	M
Voluntary education	MCRP 3-0A	M

NOTE: SECNAV = Secretary of the Navy instruction. DEERS = Defense Enrollment Eligibility Reporting System.

and respect and will maintain a professional work environment free from sexual harassment." Annual training is required for all marines and civilian personnel under the program.

The Marine Corps substance abuse training program is part of the Marine Corps Semper Fit program. Semper Fit is a comprehensive wellness program for all marines designed to "foster peak performance through maximal health" (p. vii). The goal of the substance abuse awareness program is "to provide quality, effective alcohol and substance abuse prevention programs" in which effectiveness is measured by monitoring the number of alcohol and drug incidents and "the number of Marines who report on the survey [that they are alcohol abusers or drug users or] that their job performance was affected" (p. 1-7). Annual training is required as a part of this program.

The Marine Corps program to safeguard PII requires annual training as well. The program is outlined in Marine Corps Bulletin (MCBUL) 5239, "Interim Guidance for Handling, Safeguarding, and Reporting Breaches of Personally Identifiable Information." The intent of the program is to both safeguard and maintain the public's trust regarding private information entrusted to the Marine Corps.

Marine Corps CoC training is directed by executive order and by DoD. The training educates marines on appropriate conduct in combat or if captured. It is required upon initial entry into military service and continuously throughout a marine's military career.

Marine Corps suicide awareness training is also part of the Semper Fit program. The objective of the training program is to reduce the number of suicides in the Marine Corps. Annual training is required to deliver awareness and prevention information. The Marine Corps states that the success of the program is measured by reductions in the number of suicides annually.

The Marine Corps EO and human relations training program provides different types of training and education at different levels of a marine's career. For example, one part of this training is a review of EO policy, the benefits of unbiased camaraderie and teamwork among marines, and the negative impacts of discrimination and prejudice. All Marine Corps personnel are required to receive one-hour policy review training annually.

Guidance

Marine Corps ancillary training is required by law (e.g., U.S. Code, executive orders, federal regulations) and DoD, U.S. Department of the Navy, and Marine Corps policies. For example, under the training topic "suicide awareness," we identified DoDD 6200.04 as the higher-level requirement that states that the DoD components "shall promote and improve the health of the force through programs . . . on suicide prevention" (DoDD 6200.04). MCO P1700.29 further refines this requirement and mandates that Marine Corps commands "ensure that all Marines receive education and training annually in Suicide Awareness" (MCO P1700.29)

Delivery

Like the other services, the Marine Corps employs several different delivery methods for its ancillary training programs. The methods employed are a stand-up in-person group session or an interactive CBT session.

CBT is delivered as part of the Marine Corps's larger online training platform, MarineNet. All Marine Corps ancillary online training is interactive because it requires users to respond to questions in order to navigate and complete the training. Some Marine Corps training topics, such as terrorism awareness, require a proctored online test in order to complete the course of training. To access and complete a proctored online test, an authorized proctor must log in with his or her user name and password after the login of person being tested. The proctor is essentially certifying that he or she verified the identity of the test taker and will be monitoring the fair and honest completion of the test.

Some courses, such as Marine Corps common skills, are delivered as stand-up training but have a completion test that is administered online at a later time. In addition, MarineNet hosts courses produced by other DoD agencies. The IA course is produced by DISA and is mandated for use by all DoD services. Second, rather than expend funds to develop a parallel training, the Marine Corps simply copied two Army-produced courses—law of land warfare and OPSEC—into MarineNet. Marine training officials indicated that having MarineNet as an ancillary training option is desirable because it simplifies the process of training individuals who may have missed a stand-up training session. But these same officials indicated that requiring that all ancillary training be completed online was not preferred because unit commanders typically prefer to use stand-up delivery as an opportunity to exercise leadership and mentorship with their subordinates.

Tracking

The Marine Corps has two methods for tracking training completion. One method for tracking training is the unit diary. The unit diary is a system that a unit commander uses to track and report the administrative activities of his or her unit, including such items as training or personnel arrivals and departures. In conjunction with the unit diary, marine training officials indicated that the most common process for tracking training completion is spreadsheets. These spreadsheets are then manually consolidated and rereported to higher administrative levels. To alleviate this administrative burden, the Marine Corps is implementing and deploying an online database for monitoring completion of training requirements. This system is called the Marine Corps Training Information Management System (MCTIMS). MCTIMS is a relatively new web-based application that is intended to be an authoritative data source allowing for unit training readiness assessments.

Periodicity

According to our examination of Marine Corps policies, Marine Corps ancillary training occurs at intervals from as little as a week (for the MCI study test) to five years (for combat water survival for certain categories of personnel).

Reserve Component Perspectives

Marine Corps reservists are subject to time constraints that are similar to those of other RCs, i.e., time is scarce. Marine Corps reservists must invest a significant amount of time for "formal" qualifications (e.g., rifle, pistol, water survival), on top of unit- or vocation-specific training time. By Marine Corps policy, the third priority behind the formal and vocational requirements is ancillary training (MCRP 3-0A).

An informal accounting of the time it takes to complete GMT requirements versus available man-hours indicated that 18 percent (61 of 337 hours) of a reserve marine's time could be spent on GMT requirements (Olson, 2009). Using the 1800-series MOS as an example, the estimated training time required for vocational proficiency was 60 days per year.[2] When the total training demands are considered, it becomes apparent that a change either in the amount of training or in the time available for training is needed for an RC 1800 MOS to achieve full compliance with Marine Corps training policies. The training demands are greater than the time available to complete them.

Another issue for Marine Corps reservists is access to computers for distance or online training. As the Marine Corps begins to include more ancillary training topics on MarineNet, this shortfall will become more acute. Reserve officials voiced concerns about this shortfall while continuing to stress the value of providing a maximum amount of flexibility with regard to delivery options. As an example, such flexibility allows commanders to conduct online makeup sessions for those absent from group training sessions. Alternatively, some reserve units prefer to have their members complete their ancillary training at home and online so that they can maximize the use of their drill weekends for higher-priority training.

---

[2] 1800-series MOS refers to a basic tank and assault amphibious vehicle crewman.

APPENDIX G

Coast Guard Mandated Training

Background

The Coast Guard recently renamed its GMT program to *mandated training*. With that change, the term *mandated training* now applies to training given to all Coast Guard members, both military (active and reserve) and civilian (employee and auxiliary volunteers). In the Coast Guard's "Performance, Training and Education Manual" (Commandant Instruction [COMDTINST] 1500.10C), the Coast Guard defines *mandated training* as

> A broad category of training which is needed to introduce, reinforce or clarify the law to improve the function of the federal government or protect the safety of its people; influence the behaviors and/or attitudes of the workforce towards CG Core values; or is critical to improve the function of the Coast Guard or protect the safety of its people.

The Coast Guard has transitioned from using an instruction as the authority designating the mandated training to using its learning portal as the authority mandating training.

Objectives

The Coast Guard's mandated training program is part of its larger Human Performance System (HPS). The mandated training program shares its objectives with the HPS. The objectives of the HPS are to (1) use standardized and systematic methods, (2) respond to emerging needs by working with other programs, (3) leverage technology to achieve efficiency, (4) develop the intellectual capital of the workforce, and (5) manage and allocate resources to organizational priorities. The Coast Guard's training and education manual provides a detailed outline of the HPS and its connection to mandated training.

Curriculum

The Coast Guard requires seven MT topics (U.S. Coast Guard, date unknown). It recently reduced its MT topics from 12 to seven by consolidating topics and by moving some topics from "all hands" to smaller audiences, such as "supervisors only." All current topics are listed in Table G.1. Sexual harassment and prevention (SHP) was combined with SAPR to become SHAPR. In addition, four security topics (information security management, AT/FP, OPSEC

Table G.1
Coast Guard Training Topics, Training Guidance, and Periodicity Requirements

Topic	Highest Directive	Periodicity
Information security systems	5 U.S.C. § 4118	A
No Fear Act[a]	Pub. L. 107-174 §202(c)	B
Privacy awareness	5 U.S.C. § 552a	A
SAFE	COMDTINST 5370.1A	T
SETA[b]	Executive Order 12958, COMDTINST M5510.24, COMDTINST M5530.1C	A
SHAPR	Pub. L. 108-375 §§577(b)(2), (e)(2)(A) and (F), (f)(2)(C) and (D)	A
Suicide prevention	COMDTINST 1734.1	A

NOTE: In the "Periodicity" column, A = annual (trained annually); B = biannual (trained once every two years); T = triennial (Trained once every three years). SAFE = substance abuse–free environment.

[a] The No Fear Act is whistleblower protection and is required for civilians and their supervisors.

[b] SETA is a combination of courses: (1) AT, (2) information security management, (3) OPSEC awareness, and (4) security awareness.

awareness, and security awareness) were combined into a single course titled security education and training awareness (SETA).

Of the Coast Guard's seven MT topics, all except the No Fear Act fall under the RAND definition of *GMT.*

The Coast Guard SHAPR training program is designed to foster a climate of "mutual respect and trust . . . and affirm the contributions of all personnel." It is required annually for all personnel.

The Coast Guard's information security systems (ISS) training is required annually of all personnel who operate any information technology (IT) system that handles sensitive or classified information. Because all Coast Guard personnel are expected to use Coast Guard computers in the normal course of their duties, this requirement applies to all Coast Guard personnel. ISS training is designed to educate personnel on the appropriate use of information systems and ensure that each member does his or her part to protect the integrity of the entire system.

The Coast Guard's privacy awareness training program is designed to teach personnel how to properly handle and disseminate (when appropriate) personal information held by the Coast Guard. All Coast Guard personnel are required to be trained annually.

The Coast Guard's suicide prevention training program is designed to reduce or eliminate suicides among Coast Guard personnel. All Coast Guard personnel are required to complete this training annually.

Coast Guard SAFE awareness training is designed to provide a basic awareness of the Coast Guard's SAFE program. It covers signs of abuse and explains the consequences both for the individual and for unit readiness. This training is required to be completed once every three years by all Coast Guard personnel.

The Coast Guard's SETA is a basic overview of security principles. It is designed to teach Coast Guard members how to protect themselves, their units, and the Coast Guard from individuals with malicious intent. The training is a combination of what once were four separate training topics: AT, information security management, OPSEC awareness, and security awareness. All Coast Guard members are required to complete this training annually.

Guidance

Of the seven training topics mandated by the Coast Guard, a higher authority mandates five. Coast Guard mandated training is driven by requirements in law (e.g., U.S. Code, executive orders, federal regulations), DHS policies, and Coast Guard policies. Although the Coast Guard is a part of DHS, legal provisions exist to transfer the Coast Guard to the Department of the Navy in time of war or as the President may direct. As such, the Coast Guard maintains close alignment with DoD requirements. For example, under the training topic "privacy awareness," we identified the Privacy Act of 1974 (5 U.S.C. § 552) as the legislative requirement requiring agencies that maintain systems of records to establish rules of conduct and instruct their personnel. The White House's Office of Management and Budget (OMB) further refines this in OMB Circular A-130, which states that agencies will "train their personnel in skills appropriate to the management of information." The Coast Guard policy that implements this requirement is Commandant Note 1500, which lists "Privacy Awareness: FOIA" as an annually mandated training topic.

Delivery

The Coast Guard has put considerable effort into moving all its mandated training topics to its learning portal. Although training may be completed in a classroom (i.e., stand-up) mode, the preferred method is online interactive CBT for all topics. All Coast Guard CBT is interactive because it requires the user to answer questions as he or she proceeds through the training. In addition, all topics require satisfactory completion of a test to complete the course.

The Coast Guard redesigned its online courses to present a customized experience based on a pretest. If a satisfactory score is achieved on the pretest, the training topic will be recorded as completed, and the user will receive no further training on that topic. If an unsatisfactory score is achieved on a particular section or sections of the pretest, the user is directed to a customized version of the training designed to improve his or her knowledge of those specific topics in which the scores were unsatisfactory. If new content is added to an individual mandated training curriculum, the system can be tailored to either require a user to take only that section of the training or give the user the opportunity to test out of the training. For example, the Coast Guard expects the average time for a service member to complete all seven topics in the redesigned format to be four hours. However, a Coast Guard member with 15 years of experience completed all MT in less than two hours by "testing out" of much of the subject matter.

Tracking

Once a user completes the training in the learning portal, a record of the completed training is automatically recorded in the Training Management Tool (TMT) database. For stand-up delivery, the training officer updates the TMT database manually. One unique feature of the Coast Guard's tracking system is its interface with the Coast Guard Business Intelligence (CGBI) system. This system uses a red-yellow-green stoplight design to provide each Coast Guard user with an overview of his or her readiness or compliance with service mandates, such as physical examinations, security clearance paperwork, or annual mandated training requirements. It also allows any Coast Guard computer user to easily "click through" and examine the compliance level of his or her unit or division or other units or divisions. The ease, visibility, and transparency with which the CGBI system aggregates and displays training completion rates assisted one large unit's training officer with increasing his unit's compliance with mandated training requirements from 39 percent in 2007 to 82 percent in 2008.

Periodicity

Commandant Note 1500 identifies the periodicity of the seven training topics. Five are annual, one biannual, and one triannual. The CGBI tracking system reports individualized due dates on training for an individual. For annual training, the training generally expires on December 31 of the calendar year following the last date of training. When this occurs, the user is given a red stoplight indicator on his or her personal readiness page indicating that this training is due and must be retaken. One exception to this is the ISS training, for which the training is required 365 days after the last completion date.

Reserve Component Perspectives

Coast Guard reservists are subject to the same time constraints discussed previously for the other services. Coast Guard reservists often perform their drills in direct support of their assigned units, so their training time competes directly with time devoted to mission accomplishment. The Coast Guard addresses this by providing a maximum amount of flexibility in when and where a reservist may accomplish his or her mandated training. Because the Coast Guard's learning portal is the preferred method for completing mandated training and the portal is accessible anywhere on the Internet via username and password, Coast Guard reservists have 24/7 access to mandated training courses similar to their active-duty counterparts. The Coast Guard authorized all reservists to complete mandated training as distance learning subject to written permission from their chain of command (COMDTINST M1500.10C). Completion of all mandated training courses by a reservist is equivalent to one inactive duty for training drill. To receive approval for completing the drill, the reservist must provide completion certificates to his or her chain of command.

Bibliography

AFI 36-2201—*See* U.S. Department of the Air Force, 2011.

Antiterrorism Level I Training System, undated home page.

AR 40-3—*See* U.S. Department of the Army, 2008a.

AR 350-1—*See* U.S. Department of the Army, 2007c.

AR 385-10—*See* U.S. Department of the Army, 2011.

AR 600-20—*See* U.S. Department of the Army, 2008b.

Blascovich, James J., and Christine R. Hartel, *Human Behavior in Military Contexts*, Washington, D.C.: National Academies Press, 2008.

Bush, George H. W., "Principles of Ethical Conduct for Government Officers and Employees," Washington, D.C.: White House, Executive Order 12674, April 12, 1989, as modified by Executive Order 12731, October 17, 1990.

Carter, James Earl Jr., "Occupational Safety and Health Programs for Federal Employees," Washington, D.C.: White House, Executive Order 12196, February 26, 1980. As of March 13, 2012: http://www.archives.gov/federal-register/codification/executive-order/12196.html

Clinton, William J., "Classified National Security Information," Washington, D.C.: White House, Executive Order 12958, April 17, 1995a. As of March 13, 2012: http://www.gpo.gov/fdsys/pkg/FR-1995-04-20/pdf/95-9941.pdf

———, "Access to Classified Information," Washington, D.C.: White House, Executive Order 12968, August 2, 1995b. As of March 13, 2012: http://www.gpo.gov/fdsys/pkg/FR-1995-08-07/pdf/95-19654.pdf

Codone, Susan, "Measuring Quality in the Production of Web-Based Training: Instructional Design, Process Control, and User Satisfaction," in *The Interservice/Industry Training, Simulation and Education Conference (I/ITSEC)*, Vol. 2001, 2001, pp. 3–4.

COMDTINST 1734.1—*See* U.S. Coast Guard, 1997.

COMDTINST 5370.1A—*See* U.S. Cost Guard, 1999.

COMDTINST M1500.10C—*See* U.S. Coast Guard, 2009b.

COMDTINST M5510.24—*See* U.S. Coast Guard, 2005a.

COMDTINST M5530.1C—*See* U.S. Coast Guard, 2001b.

DA PAM 350-38—*See* U.S. Department of the Army, 2009b.

DoDD 1010.4—*See* U.S. Department of Defense, 1997e.

DoDD 1020.02—*See* U.S. Department of Defense, 2009b.

Eisenhower, Dwight D., "Code of Conduct for Members of the Armed Forces of the United States," Washington, D.C.: White House, Executive Order 10631, August 17, 1955. As of March 13, 2012: http://www.archives.gov/federal-register/codification/executive-order/10631.html

Executive Order 10631—*See* Eisenhower, 1955.

FM 3-25.150—*See* U.S. Department of the Army, 2002.

FM 21-20—*See* U.S. Department of the Army, 1998.

Kulik, James A., "Meta-Analytic Studies of Findings on Computer-Based Instruction," in Eva L. Baker and Harold F. O'Neil Jr., eds., *Technology Assessment in Education and Training*, Hillsdale, N.J.: L. Earlbaum Associates, 1994, pp. 9–33.

MCBUL 5239—*See* U.S. Marine Corps, 2008.

MCO 1500.52C—*See* U.S. Marine Corps, 2004c.

MCO 1510.121A—*See* U.S. Marine Corps, 2004d.

MCO 1553.3A—*See* U.S. Marine Corps, 2004a.

MCO 1741.11C—*See* U.S. Marine Corps, 2005.

MCO 3400.3F—*See* U.S. Marine Corps, 2004b.

MCO 3574.2K—*See* U.S. Marine Corps, 2007.

MCO 4600.40A—*See* U.S. Marine Corps, 2002.

MCRP 3-0A—*See* U.S. Marine Corps, 1996.

National Security Council, "Reporting Hostile Contacts and Security Awareness," National Security Decision Directive 197, November 1, 1985. As of March 13, 2012:
http://www.fas.org/irp/offdocs/nsdd/nsdd-197.htm

NAVADMIN 033/09—*See* U.S. Department of the Navy, 2009a.

NAVADMIN 098/10—*See* U.S. Department of the Navy, 2010.

NSC-NSDD-197—*See* National Security Council, 1985.

Office of Government Ethics, "Executive Agency Ethics Training Programs, Generally," *Code of Federal Regulations*, Title 5, Part 2638, Section 2638.701, RIN 3209-AA07, September 18, 1990.

Office of Management and Budget, "Management of Federal Information Resources," Circular A-130, February 8, 1996. As of March 13, 2012:
http://www.whitehouse.gov/omb/circulars_a130

Olson, Taz, "Simply Not Enough Time: Time Management Is Needed," *Marine Corps Gazette*, Vol. 93, No. 10, October 2009, pp. 54–58.

OPNAVINST 1500.22E—*See* U.S. Department of the Navy, 2000.

Public Law 107-174, Notification and Federal Employee Antidiscrimination and Retaliation Act of 2002, May 15, 2002.

Public Law 107-347, E-Government Act of 2002, December 17, 2002.

Public Law 108-375, Ronald W. Reagan National Defense Authorization Act for Fiscal Year 2005, October 28, 2004.

SECNAV 1740.4—*See* U.S. Department of the Navy, 2007e.

Shanley, Michael G., Matthew W. Lewis, and Susan G. Straus, Jeff Rothenberg, and Lindsay Daugherty, *The Prospects for Increasing the Reuse of Digital Training Content*, Santa Monica, Calif.: RAND Corporation, MG-732-OSD, 2009. As of January 16, 2011:
http://www.rand.org/pubs/monographs/MG732.html

Sitzmann, Traci, Kurt Kraiger, David Stewart, and Robert Wisher, "The Comparative Effectiveness of Web-Based and Classroom Instruction: A Meta-Analysis," *Personnel Psychology*, Vol. 59, 2006, pp. 623–664.

Society of American Military Engineers, "Joint Engineer Training," presented at the Third Annual Joint Engineer Noncommissioned Officer Symposium, Fort Leonard Wood, Mo., October 3, 2006. As of January 31, 2012:
http://www.flwsame.com/Joint%20Engineer%20Training.ppt

Straus, Susan G., Jolene Galegher, Michael G. Shanley, and Joy S. Moini, *Improving the Effectiveness of Distributed Learning: A Research and Policy Agenda*, Santa Monica, Calif.: RAND Corporation, OP-156-A, 2006. As of January 16, 2011:
http://www.rand.org/pubs/occasional_papers/OP156.html

Straus, Susan G., Michael G. Shanley, Rachel M. Burns, Anisah Waite, and James C. Crowley, *Improving the Army's Assessment of Interactive Multimedia Instruction Courseware*, Santa Monica, Calif.: RAND Corporation, MG-865-A, 2009. As of March 13, 2012:
http://www.rand.org/pubs/monographs/MG865.html

"The Lautenberg Amendment," undated. As of April 16, 2012:
http://www.monterey.army.mil/legal/military_law/lautenberg_amendment.pdf

Thompson, Tonnette, "Ancillary Training Cut from Nine Blocks to Three," Sheppard Air Force Base, Texas, November 9, 2007. As of March 20, 2012:
http://www.sheppard.af.mil/news/story.asp?id=123075434

U.S. Army Distributed Learning System, "What Is DLS?" undated home page. As of March 20, 2012:
https://www.dls.army.mil/

U.S. Coast Guard, "General Mandated Training Requirements," ALCOAST 127-08, Commandant Note 1500, date unknown.

———, "The Coast Guard Freedom of Information and Privacy Acts Manual," Commandant Instruction M5260.3, June 14, 1993.

———, "Suicide Prevention," Commandant Instruction 1734.1, February 28, 1997.

———, "Workplace Violence and Threatening Behavior," Commandant Instruction 5370.1A, October 29, 1999.

———, "Classified Information Management Program," Commandant Instruction Manual 5510.23, January 17, 2001a.

———, "Physical Security and Force Protection Program," Commandant Instruction M5530.1C, December 17, 2001b.

———, "Operations Security Program," Commandant Instruction M5510.24, August 10, 2005a.

———, "Equal Opportunity Manual," Commandant Instruction Manual 5350.4B, November 1, 2005b.

———, "Sexual Assault Prevention and Response Program," Commandant Instruction 1754.10C, December 20, 2007.

———, "Mandated ISS Web-Based User Awareness Availability," ALCOAST 176/09, March 25, 2009a.

———, "Performance, Training and Education Manual," Commandant Instruction M1500.10C, May 2009b. As of March 22, 2012:
http://www.uscg.mil/directives/cim/1000-1999/CIM_1500_10C.pdf

U.S. Code, Title 5, Government Organization and Employees, Section 552a, Privacy and Social Security Numbers, December 31, 1974.

———, Title 5, Part 930, Programs for Specific Positions and Examinations, January 5, 2009.

———, Title 5, Government Organization and Employees, Part III, Employees, Subpart C, Employee Performance, Chapter 41, Training, Section 4118, Regulations.

———, Title 10, Article 137, Uniform Code of Military Justice, January 5, 2009.

———, Title 10, Section 654, Policy Concerning Homosexuality in the Armed Forces, January 9, 2009.

———, Title 18, Part 1, Chapter 113C, Torture, January 8, 2008.

———, Title 44, Section 2904, General Responsibilities for Records Management, January 2003.

U.S. Department of the Air Force, "Service to Survivors," Department of the Air Force Policy Directive 34-11, May 1, 2000a.

———, "Counterintelligence," Department of the Air Force Instruction 71-101, August 1, 2000b.

———, "Alcohol and Drug Abuse Prevention and Treatment Program," Department of the Air Force Instruction 44-121, September 26, 2001.

———, "Suicide and Violence Prevention," Department of the Air Force Instruction 44-154, January 3, 2003.

———, "Privacy Act Program," Department of the Air Force Instruction 33-332, January 29, 2004a.

———, "Applying North Atlantic Treaty Organization Protection Standards," Department of the Air Force Instruction 31-406, July 29, 2004b.

———, "Assistance to Survivors of Persons Killed in Air Force Aviation Mishaps and Other Incidents," Department of the Air Force Instruction 34-1101, November 1, 2006.

———, "Air Force Emergency Management Program Planning and Operations," Department of the Air Force Instruction 10-2501, January 24, 2007a.

———, "The U.S. Air Force Traffic Safety Program," Department of the Air Force Instruction 91-207, May 11, 2007b.

———, "Operations Security," Department of the Air Force Instruction 10-701, October 18, 2007c.

———, "Administration of Military Justice," Department of the Air Force Instruction 51-201, December 21, 2007d.

———, "Sexual Assault Prevention and Response Program," Department of the Air Force Policy Directive 36-60, March 28, 2008a.

———, "Information Assurance (IA) Management," Department of the Air Force Instruction 33-200, December 23, 2008b.

———, "Combating Trafficking in Persons," Department of the Air Force Instruction 36-2921, February 23, 2009a.

———, "Antiterrorism," Department of the Air Force Instruction 10-245, March 30, 2009b.

———, "Air Force Training Program," Department of the Air Force Instruction 36-2201, September 15, 2010, incorporating change 1, March 8, 2011. As of April 16, 2012:
http://www.e-publishing.af.mil/shared/media/epubs/AFI36-2201.pdf

U.S. Department of the Army, "Prevention of Motor Vehicle Accidents," Army Regulation 358-55, March 12, 1987.

———, "U.S. Army SAEDA," Army Regulation 381-12, January 15, 1993.

———, "Physical Fitness Training," Field Manual 21-20, October 1, 1998. As of March 21, 2012:
http://www.apft.net/FM21-20.pdf

———, "The Army Public Affairs Program," Army Regulation 360-81, September 15, 2000.

———, "Combatives," Field Manual 3-25.150, January 18, 2002.

———, "Family Support, Child Custody, and Paternity," Army Regulation 608-99, October 29, 2003.

———, "Military Justice," Army Regulation 27-10, November 16, 2005.

———, "Preventative Medicine," Army Regulation 40-5, May 25, 2007a.

———, "Active and Reserve Components Enlistment Program," Army Regulation 601-210, June 1, 2007b.

———, "Army Training and Leader Development," Army Regulation 350-1, August 3, 2007c.

———, "Medical, Dental, and Veterinary Care," Army Regulation 40-3, February 22, 2008a.

———, "Army Command Policy," Army Regulation 600-20, March 18, 2008b.

———, "The Army Substance Abuse Program," Army Regulation 600-85, February 2, 2009a.

———, "Standards in Training Commission," Department of the Army Pamphlet 350-38, May 13, 2009b.

———, "The Army Safety Program," Army Regulation 385-10, October 4, 2011. As of March 21, 2012:
http://www.apd.army.mil/pdffiles/r385_10.pdf

U.S. Department of Defense, "Distribution Statements on Technical Documents," Department of Defense Directive 5320.24, date unknown.

———, "Withholding of Unclassified Technical Data from Public Disclosure," Department of Defense Directive 5230.25, November 6, 1984.

———, "Personnel Security Program Regulation," Department of Defense Directive 5200.2-R, January 1987.

———, "Department of Defense Civilian Employee Drug Abuse Testing Program," Department of Defense Directive 1010.9, August 23, 1988.

———, "Family Policy," Department of Defense Directive 1342.17, December 30, 1988.

———, "Department of Defense Military Equal Opportunity Program," Department of Defense Directive 1350.2, August 18, 1995.

———, "Personnel Security Program," Department of Defense Directive 5200.2-R, February 23, 1996a.

———, "Civilian Employment and Reemployment Rights of Applicants for, and Service Members and Former Service Members of the Uniformed Services," Department of Defense Instruction 1205.12, April 4, 1996b.

———, "Information Security Program," Department of Defense 5200.1-R, January 1997a. As of March 14, 2012:
http://www.fas.org/irp/doddir/dod/5200-1r.pdf

———, "Department of Defense Information Security Program," Department of Defense Directive 5200.1, January 14, 1997b.

———, "Department of Defense Newspapers, Magazines and Civilian Enterprise Publications," Department of Defense Instruction 5120.4, July 16, 1997c.

———, "Department of Defense Civilian Employee Drug Abuse Testing Program," Department of Defense Directive 1010.9, September 3, 1997d.

———, "Drug and Alcohol Abuse by Department of Defense Personnel," Department of Defense Directive 1010.4, September 3, 1997e.

———, "Department of Defense Safety and Occupational Health Program," Department of Defense Instruction 6055.1, August 19, 1998.

———, "Development of Interactive Multimedia Instruction," MIL-HDBK-29612-3, July 30, 1999.

———, "Records Management Program," Department of Defense Directive 5015.2, March 6, 2000a.

———, "Training and Education to Support the Code of Conduct," Department of Defense Directive 1300.7, December 8, 2000b.

———, "Armed Forces Community Relations," Department of Defense Instruction 5410.19, November 13, 2001.

———, "Department of Defense Physical Fitness and Body Fat Program," Department of Defense Instruction 1308.3, November 5, 2002a.

———, "Department of Defense Policy for Implementation of Domestic Violence Misdemeanor Amendment to the Gun Control Act for Military Personnel," Personnel and Readiness Memorandum, November 27, 2002b.

———, "Information Assurance Implementation," Department of Defense Instruction 8500.2, February 6, 2003a.

———, "Survivor Annuity Programs for the Uniformed Services," Department of Defense Directive 1332.27, June 26, 2003b.

———, "Department of Defense Civilian Equal Employment Opportunity Program," Department of Defense Directive 1440.1, May 21, 1987, certified current as of November 21, 2003c.

———, "Department of Defense Physical Fitness and Body Fat Program," Department of Defense Directive 1308.1, June 30, 2004a.

———, "Counterintelligence (CI) Awareness, Briefing, and Reporting Programs," Department of Defense Instruction 5240.6, August 7, 2004b.

———, "Force Health Protection," Department of Defense Directive 6200.04, October 9, 2004c.

———, "Personal Financial Management for Service Members," Department of Defense Instruction 1342.27, November 12, 2004d.

———, "Department of Defense Military Equal Opportunity Program," Department of Defense Directive 1350.2, August 18, 2005a.

———, "Qualification Standards for Enlistment, Appointment, and Induction," Department of Defense Instruction 1304.26, September 20, 2005b.

———, "Department of Defense Operations Security Program," Department of Defense Directive 5205.02, March 6, 2006a.

———, "Department of Defense Law of War Program," Department of Defense Directive 2311.01, May 9, 2006b.

———, "Sexual Assault Prevention and Response Program Procedures," Department of Defense Instruction 6495.02, June 23, 2006c.

———, "Department of Defense Antiterrorism Standards," Department of Defense Instruction 2000.16, October 2, 2006d.

———, "Combating Trafficking in Persons," Department of Defense Instruction 2200.01, February 16, 2007a.

———, "Department of Defense Privacy Program," Department of Defense Directive 540.11," May 8, 2007b.

———, "Department of Defense Privacy Program," Department of Defense Directive 5400.11-R, May 14, 2007c.

———, "Annual Ethics Briefing," Joint Ethics Regulation Chapter 11, Department of Defense 5500.7-R, November 29, 2007d.

———, "Political Activities by Members of the Armed Forces on Active Duty," Department of Defense Directive 1344.10, February 19, 2008a.

———, "Clearance of Department of Defense Information for Public Release," Department of Defense Directive 5230.9, August 22, 2008b.

———, "Department of Defense Security Program and Protection of Sensitive Compartmented Information," Department of Defense Instruction 5200.1, October 9, 2008c.

———, "Private Organizations on Department of Defense Installations," Department of Defense Directive 1000.15, October 24, 2008d.

———, "DoD Operations Security (OPSEC) Program Manual," Department of Defense Manual 5205.02-M, November 3, 2008e. As of March 14, 2012:
http://www.dtic.mil/whs/directives/corres/pdf/520502m.pdf

———, "Military Training," Department of Defense Directive 1322.18, January 13, 2009a. As of March 14, 2012:
http://www.dtic.mil/whs/directives/corres/pdf/132218p.pdf

———, "Diversity Management and Equal Opportunity (EO) in the Department of Defense," Department of Defense Directive 1020.02, February 5, 2009b. As of March 14, 2012:
http://www.dtic.mil/whs/directives/corres/pdf/102002p.pdf

———, "Department of Defense Traffic Safety Program," Department of Defense Instruction 6055.04, April 20, 2009c.

———, "Survivor Annuity Program Administration" Department of Defense Instruction 1332.42, June 23, 2009d.

———, "Programs for Military Morale, Welfare, and Recreation," Department of Defense Instruction 1015.10, July 6, 2009e.

———, "DoD Law of War Program," Department of Defense Directive 2311.01E, certified current as of February 22, 2011. As of March 21, 2012:
http://www.dtic.mil/whs/directives/corres/pdf/231101e.pdf

———, "Information Assurance Workforce Improvement Program," Department of Defense Directive 8570.01-M, December 19, 2005, incorporating change 3, January 24, 2012. As of April 16, 2012:
http://www.dtic.mil/whs/directives/corres/pdf/857001m.pdf

U.S. Department of the Navy, "N@vy Knowledge Online," undated website. As of March 21, 2012:
https://wwwa.nko.navy.mil/portal/home/

———, "Drug and Alcohol Abuse Program Management for the Office of the Chief of Naval Operations and (CNO) and CNO (OP-09B) Claimancy," Office of the Chief of Naval Operations Instruction 5350.7, January 10, 1992.

———, "Operations Security," Office of the Chief of Naval Operations Instruction 3432.1, August 29, 1995.

———, "General Military Training (GMT) and Naval Military Training (NMT)," Office of the Chief of Naval Operations Instruction 1500.22E, August 1, 2000.

———, "Department of the Navy Antiterrorism/Force Protection Program," Secretary of the Navy Instruction 3300.2A, March 21, 2001.

———, "Operational Risk Management," Office of the Chief of Naval Operations Instruction 3500.39B, July 30, 2004.

———, "Department of the Navy Policy on Hazing," Secretary of the Navy Instruction 1610.2A, July 15, 2005a.

———, "Family Advocacy Program," Secretary of the Navy Instruction 1752.3B, November 10, 2005b.

———, "Sexual Assault Prevention and Response," Secretary of the Navy Instruction 1752.4a, December 1, 2005c.

———, "Department of the Navy Policy on Sexual Harassment," Secretary of the Navy Instruction 5300.26D, January 3, 2006a.

———, "Equal Opportunity within the Department of the Navy," Secretary of the Navy Instruction 5350.16a, December 18, 2006b.

———, "Health and Wellness Promotion Program," Office of the Chief of Naval Operations Instruction 6100.2A, March 15, 2007a.

———, "Personnel Procurement, Military Entrance Processing Station (MEPS)," Office of the Chief of Naval Operations Instruction 1100.4C, March 23, 2007b.

———, "Navy Fraternization Policy," Office of the Chief of Naval Operations Instruction 5370.2C, April 26, 2007c.

———, "Department of the Navy Equal Opportunity Policy," Office of the Chief of Naval Operations Instruction 5354.1F, July 25, 2007d.

———, "Department of the Navy Personal Financial Management Education, Training, and Counseling Program," Secretary of the Navy Instruction 1740.4, October 9, 2007e.

———, "Department of the Navy Core Values Charter and Ethics Training," Secretary of the Navy Instruction 5350.15C, January 31, 2008a.

———, "Department of the Navy Personal Financial Management (PFM) Education, Training, and Counseling Program," Office of the Chief of Naval Operations Instruction 1740.5, July 2, 2008b.

———, "FY-09 General Military Training (GMT)," Navy Administrative Message 033/09, January 2009a. As of March 21, 2012:
http://www.public.navy.mil/bupers-npc/reference/messages/Documents/NAVADMINS/NAV2009/NAV09033.txt

———, "Personnel Recovery," Office of the Chief of Naval Operations Instruction 1000.24C, February 12, 2009b.

———, "Navy Message dtg 171653Z APR 09, NAVADMIN 114/09," Chief of Naval Operations, April 2009c.

———, "Department of the Navy Law of War Program," Secretary of the Navy Instruction 3300.1C, May 28, 2009d.

———, "Department of the Navy Information Assurance Policy," Secretary of the Navy Instruction 5239.3B, June 17, 2009e.

———, "Management of Human Immunodeficiency Virus Infection in the Navy and Marine Corps," Secretary of the Navy Instruction 5300.3D, October 5, 2009f.

———, "General Military Training," Navy Administrative Message 098/10, March 17, 2010. As of March 19, 2012:
http://www.public.navy.mil/bupers-npc/reference/messages/Documents/NAVADMINS/NAV2010/NAV10098.txt

U.S. Marine Corps, "Marine Corps Transportation Manual," Marine Corps Order P4600.7C, October 20, 1980.

———, "Unit Training Management Guide," Marine Corps Reference Publication 3-0A, November 25, 1996.

———, "Marine Corps Semper Fit Program Manual," Marine Corps Order P1700.29, November 8, 1999.

———, "Marine Corps Traffic Safety Program," Marine Corps Order 5100.19E, December 29, 2000.

———, "Government Travel Charge Card Program," Marine Corps Order 4600.40A, May 13, 2002.

———, "Marine Corps Equal Opportunity Manual," Marine Corps Order P5354.1D, April 14, 2003.

———, "Unit Training Management (UTM)," Marine Corps Order 1553.3A, January 22, 2004a.

———, "Nuclear, Biological, and Chemical Defense Training," Marine Corps Order 3400.3F, March 1, 2004b.

———, "Marine Combat Water Survival," Marine Corps Order 1500.52C, July 31, 2004c.

———, "Marine Corps Common Skills Program," Marine Corps Order 1510.121A, October 1, 2004d.

———, "Survivor Annuity Programs," Marine Corps Order 1741.11C, February 25, 2005.

———, "Sexual Harassment," Marine Corps Order 1000.9A, May 30, 2006.

———, "Marine Corps Combat Marksmanship Programs," Marine Corps Order 3574.2K, August 1, 2007.

———, "Marine Corps Physical Fitness Program," Marine Corps Order 6100.13, August 1, 2008.

———, "Interim Guidance for Handling, Safeguarding, and Reporting Breaches of Personally Identifiable Information," Marine Corps Bulletin 5239, August 29, 2008.

Viccellio, Henry Jr., "The Joint Challenge to Interservice Training," *Joint Forces Quarterly*, Spring 1995, pp. 43–47. As of December 9, 2012:
http://handle.dtic.mil/100.2/ADA528868

Yardley, Roland, Dulani Woods, and Cesse Ip, *Standardization and Reduction of Reserve Components Ancillary Training*, Santa Monica, Calif.: RAND Corporation, November 2009, not available to the general public.